Robert Saundby

Lectures on Diabetes

Including the Bradshawe Lecture, Delivered before the Royal College...

Robert Saundby

Lectures on Diabetes

Including the Bradshawe Lecture, Delivered before the Royal College...

ISBN/EAN: 9783337171155

Printed in Europe, USA, Canada, Australia, Japan

Cover: Foto ©berggeist007 / pixelio.de

More available books at **www.hansebooks.com**

LECTURES ON DIABETES:

INCLUDING

THE BRADSHAWE LECTURE,

DELIVERED BEFORE THE ROYAL COLLEGE OF PHYSICIANS ON
AUGUST 18TH, 1890.

BY

ROBERT SAUNDBY, M.D. EDIN., F.R.C.P. LOND.,

EMERITUS SENIOR PRESIDENT OF THE ROYAL MEDICAL SOCIETY; FELLOW OF THE ROYAL
MEDICO-CHIRURGICAL SOCIETY; MEMBER OF THE PATHOLOGICAL SOCIETY OF
LONDON; PHYSICIAN TO THE GENERAL HOSPITAL; CONSULTING PHYSICIAN
TO THE EYE HOSPITAL; AND CONSULTING PHYSICIAN TO THE
HOSPITAL FOR DISEASES OF WOMEN, BIRMINGHAM.

WITH ILLUSTRATIONS.

BRISTOL: JOHN WRIGHT & CO.
LONDON: SIMPKIN, MARSHALL, HAMILTON, KENT & CO., LIMITED.

1891.

JOHN WRIGHT AND CO.
PRINTERS AND PUBLISHERS, BRISTOL.

PREFACE.

THE very kind reception accorded to my "Lectures on Bright's Disease," has encouraged me to publish the present volume.

These Lectures have, in great part, appeared already in print, but, with the exception of the Bradshawe Lecture, which is reproduced *verbatim*, they have undergone considerable modifications.

I desire to express my grateful thanks to the Right Hon. LORD KNUTSFORD, G.C.M.G., Her Majesty's Secretary of State for the Colonies, for the valuable assistance he has afforded me by obtaining for me a return showing the incidence of Diabetes Mellitus in the British Colonies.

I wish to acknowledge my indebtedness to many successive pathologists, resident medical officers and clinical clerks at the General Hospital, for their indispensable co-operation in these pathological and clinical studies, and I desire to thank my friend, Mr. GILBERT BARLING, for his kindness in helping me to see this book through the press.

BIRMINGHAM,
    January, 1891.

CONTENTS.

Part I.—DIABETES MELLITUS.

CHAP.		PAGES
I.	INTRODUCTION—HISTORY—PHYSIOLOGY OF GLYCOSURIA, ETC.	1—19
II.	ETIOLOGY	20—53
III.	MORBID ANATOMY—THE BRADSHAWE LECTURE	54—78
IV.	CLINICAL HISTORY	79—133
V.	DIABETIC COMA	134—180
VI.	TREATMENT	181—204

Part II.—DIABETES INSIPIDUS.

VII.	ETIOLOGY—MORBID ANATOMY—CLINICAL HISTORY—TREATMENT	205—219
	GENERAL INDEX	220—232

LIST OF ILLUSTRATIONS.

	PAGE
HEART MUSCLE, FROM A CASE OF VERY ADVANCED FATTY AND FIBROID DEGENERATION	60
HEART MUSCLE, APPARENTLY HEALTHY, EXCEPT FOR A FEW GRANULES OF GLYCOGEN	61
SECTION OF LIVER SHEWING COMMENCING INTESTINAL HEPATITIS	65
SECTION OF ATROPHIC CIRRHOTIC PANCREAS	68
HIGHLY MAGNIFIED VIEW OF A GROUP OF PANCREATIC ACINI	69
SECTION OF ENLARGED PANCREAS SHEWING COMMENCING CIRRHOSIS	70
PORTION OF HENLE'S LOOPED TUBE, SHEWING HYALINE DEGENERATION OF EPITHELIUM	72
RENAL TUBULE, SHEWING MASSES OF GLYCOGEN	73
PORTION OF RENAL TUBULE, SHEWING FAT DROPLETS	74
GERRARD'S GLYCOSOMETER	90

LECTURES ON DIABETES.

PART I.—DIABETES MELLITUS.

CHAPTER I.

DIABETES Mellitus is essentially one of those rarer diseases which can only be effectively studied, at least in this country, by those who reside in great centres of population and have the extensive practice of a large hospital from which to draw their cases, as well as the opportunities of treating numerous individuals suffering from this malady under those conditions of control which can rarely be obtained except in the wards of such an institution. In the following lectures, it is my purpose to describe this disease in the light of the clinical and pathological experience acquired in fifteen years of hospital study, and I shall try to infuse into them as much as is possible of the results of my bedside observations and *post mortem* examinations, testing all theories by these standards. There is probably no disease which is so overweighted by theoretical considerations derived from an overwhelming amount of physiological experiment. But it is absolutely necessary to keep such physiological considerations in their proper place, using them only for the purpose of elucidating the facts of disease, not substituting one for the other, as if a frog with its spinal cord divided or a rabbit with its medulla injured were really and truly a case of Diabetes.

HISTORY.

Hirsch can hardly be held to be free from a tendency to exaggeration when he says that the history of diabetes goes back to the most remote antiquity. He is doubtless right in quoting the passages from the *Ayur Veda* in proof that the disease was known long ago to the Hindoos, but this *Veda* was certainly not written till the sixth century of our era, and is supposed to have been in great part a copy of Greek medicine transmitted through Arab sources. It is however very noteworthy that in these quotations the urine is described as *sweet*, and in a Cingalese treatise of the fifteenth century the disease is referred to as a "Madu mehe" or "honey urine," though no mention of this very singular peculiarity can be traced among European writers until two centuries later. No mention of diabetes can be discovered in the writings of Hippocrates; and though Celsus described a disease attended by polyuria, wasting and bodily illness, the term diabetes was first used by Aretæus (A.D. *circa* 150). Galen wrote on the disease at some length. But neither of these authors referred to the *sweetness* of the urine, and this important fact remained undescribed until it was noted by Willis (1679). Sydenham, contemporary with Willis, wrote a good account of diabetes in which he drew attention to the importance of abstinence from vegetable food. "Let the patient," he wrote, "eat food of easy digestion, such as veal, mutton and the like, and *abstain from all sorts of fruit and garden stuff*."

A century later Dobson (1776) evaporated two quarts of diabetic urine and obtained a cake which weighed 4oz., 2dr., 2scr. This cake "smelt sweet, like brown sugar, and could not be distinguished from sugar, except that the sweetness left a slight sense of coolness on the palate," probably due to the presence of a certain proportion of sodium chloride.

Cullen in the first edition of his Practice of Physic (1784) wrote that the urine in diabetes contained a considerable quantity of a saccharine matter which seemed to be exactly of the nature of sugar. So impressed was he with this discovery that he would hardly admit the existence of a non-saccharine variety (diabetes insipidus), though he thought he had seen a case.

The publication of Rollo's excellent cases (1797) marks an era in the history of this disease, because of his powerful advocacy of meat diet, and the success which attended the method of treatment he employed, so that he may be said to have established the foundation of our modern practice.

Latham (1810) went a step farther, recognising two types of diabetes, one *saccharine*, and the other *serous*, and combining Rollo's animal diet with such excellent remedies as opium, iron and alkalies.

Gregory, in the appendix to his edition of Cullen's Practice of Physic (1829), recognised two forms, D. insipidus and D. mellitus. He wrote slightingly of Rollo's treatment which seems to have suffered a check in its popularity, when it became known that in very many cases it failed to cure the disease. Nevertheless this method held its ground until modified by Bouchardat (1841) who introduced gluten bread.

In 1848 Claude Bernard discovered sugar in the liver after death, and in 1859 he proved that this sugar existed in the liver during life in the form of glycogen. He believed that glycogen was converted into sugar by the action of a diastatic ferment contained in the blood. In 1858 Pavy proved that the formation of sugar in the liver in such large quantities was a *post mortem* phenomenon, and he disputed the conversion of glycogen into sugar at all as a physiological process, suggesting that it really left the liver in some other form, such as fat. The great discovery of Bernard that puncture of the floor of the fourth ventricle was followed by glycosuria, gave, however, a possible explanation of the pathology of diabetes, though this has been greatly diminished in value by the subsequent experiments which have shewn that lesions of many parts of the nervous system are followed by the same phenomenon. The present position of the problem cannot be described adequately by continuing the historical method. In order to render it as clear as possible the physiology of glycosuria will be explained at length in the next section, while the etiology, morbid anatomy and pathology of diabetes will be dealt with separately in subsequent pages.

THE PHYSIOLOGY OF GLYCOSURIA.

By glycosuria is understood the presence in the urine of glucose ($C_6H_{12}O_6$). That this substance is present in small quantities in normal urine has been long known, but in such small amount that it can only be demonstrated by the usual tests after concentrating a large bulk of urine. Recently Wedenski has utilised Baumann's discovery, that benzoyl chloride forms insoluble compounds with carbo-hydrates, in order to demonstrate the existence of this physiological glycosuria.

The *glucose* found in the urine whether normally or under pathological conditions is as a rule grape sugar (dextrose). *Lævulose* or intestine sugar is stated to occur rarely, but is not distinguished from it by the ordinary tests. *Inosite* or muscle sugar is an isomer of glucose, but occurs very rarely in the urine, and is of little practical importance; it does not reduce cupric salts.

Lactose, ($C_{12}H_{22}O_{11}$) or milk sugar, is a *sucrose*, but it is sometimes present in the urine during pregnancy or lactation, and reduces copper salts, though less actively than glucose, so that its presence is liable to be mistaken for that of the latter.

The reduction of copper in Fehling's or Trommer's test is so very generally regarded clinically as proof of glycosuria that great interest attaches to the recent discovery of the not uncommon occurrence of a non-saccharine body in the urine which possesses this power in a high degree. Schmiedeberg and Meyer have shewn that this substance is glycuronic acid. It is found in the urine after the administration of various drugs, chloral hydrate, croton chloral, camphor, phenacetin, morphia, chloroform and curare. Ashdown has met with it in the urine of a young man, aged twenty-four, who enjoyed perfect health, or at least a complete sense of well-being. Glycuronic acid ($C_6H_{10}O_7$) occurs in the urine in combination with urea, but the nature of the compound has not been definitely determined. Ashdown recommends that in all doubtful cases the urine should be fermented.

Glucose is found in minute quantities in healthy blood, chyle and muscle, but some doubt exists as to whether it is present physiologically in the liver. It is formed by the action of diastatic ferments upon other carbo-hydrates during the process of digestion. Glucose so formed is absorbed and a large part reaches the liver where it is transformed into glycogen.

Glycogen ($6(C_6H_{10}O_5) + H_2O$) is formed in the liver cells in the form of amorphous granules collected around their nuclei, and is irregularly distributed through the liver. It is soluble in water, but not readily diffusible, and gives a deep red colour with iodine solution. With diastatic ferments, or when boiled with a mineral acid it forms grape sugar. The following directions for its preparation are taken from Landois and Stirling :—" Let a rabbit have a hearty meal and kill it three or four hours thereafter. The liver is removed immediately after death; it is cut into fine pieces, plunged into boiling water and boiled for some time in order to obtain a watery extract of the liver cells. To the cold filtrate (of this extract) are added alternately dilute hydrochloric acid and potassio-mercuric iodide as long as a precipitate occurs. The albuminates or proteids are precipitated by the iodine compound in the presence of free HCL. It is then filtered, when a clear opalescent fluid, containing the glycogen in solution, is obtained. The glycogen is precipitated from the filtrate, as a white amorphous powder, on adding an excess of 70-80 p. c., alcohol. The precipitate is washed with 60 p. c., and afterwards with 95 p. c., alcohol, then with ether, and lastly with absolute alcohol ; it is dried over sulphuric acid and weighed (Brücke)."

The quantity in the liver is increased by adding starch, milk, fruit, or cane-sugar, alkalies (Dufour), glycerine, or inosite to the food, while it is diminished by a purely albuminous or purely fatty diet, and disappears during hunger. It is also diminished by cold and violent muscular exercise (Külz), and by ligature of the bile duct (Wickham Legg, von Wittich).

When rabbits are kept without food for six days, it is at least four hours before glycogen is found in their livers (Külz). It is noteworthy that in diabetes an increase of sugar can be noted in from 1 to $1\frac{1}{2}$ hours after the use of starchy food.

Glycogen is not a product peculiar to the liver; in fœtal life it is found in all the tissues of the body, and in the adult occurs in muscle, cartilage, the colourless blood corpuscles, the testicle, etc.

Glycogen can be formed directly from sugar in muscle, and by injecting syrup into the circulation a direct conversion of sugar into glycogen can be proved to take place.

As has been already stated, under physiological conditions, glycogen is formed in the liver cells from the products of digestion of starchy and saccharine food.

It is probable that in health the glycogen is very gradually converted into sugar, in very small quantities (Landois), as it is found to diminish during hunger (Pavy), exercise, and exposure to cold. The sugar so produced passes into the circulation where it is rapidly used up as it passes through the systemic capillaries.

But whenever there is derangement of the hepatic circulation, permitting a greater afflux of arterial blood to the liver the formation of sugar from glycogen is increased, and sugar in larger quantities passes into the general circulation.

This sugar production is thought to be effected by a ferment formed on the contact of the liver cells with *arterial* blood, simple venous congestion not being followed by any increase (Pavy).

The vascular derangement which permits this great afflux of arterial blood to the liver is assumed to be a vaso-motor paralysis of the splanchnic area, in consequence of which the blood reaches the portal vein without becoming deoxygenised.

Although this seems to be the most probable conclusion, the problem is by no means settled.

Kühne and Heynsius have suggested that glycocholic acid may split up into urea and glucose, and they found that the introduction of glycocin ($C_2H_5NO_2$, or sugar of gelatine, a constituent of bile) into the blood is followed by an increase in the amount of urea in the liver and urine, and of glycogen in the liver.

According to Seegen, the liver contains at the instant when life ceases 0·4 to 0·6 per cent. of sugar. (Bernard, 0·2 to 0·3 per cent.; Dalton, 0·2 to 0·4 per cent.; Pavy, 0·02 to 0·05 per cent.)

Also while carotid blood and mixed venous blood

from the right side of the heart contain about the same amount of sugar (0·109 to 0·153 per cent., a little in excess of the portal blood,) the blood of the hepatic vein contains *twice as much sugar as the portal vein.*

Bernard thought that the glycogen of the liver, formed from the starchy matter of food, is stored up and converted gradually into sugar by a ferment in the liver.

This ferment Bernard believed he had isolated, but it has since been shewn that any soluble albuminous body may be made to yield a solution capable of converting starch into sugar (Lépine), such sugar being always a glucose, while the sugar produced by true ferments (*e.g.*, ptyalin, diastase) has a slighter reducing power with a greater rotatory power and is probably identical with *maltose.* According to Dastre, the liver contains no diastatic ferment, but the liver cells possess the power of converting sucrose into glucose.

Again, according to Bernard, as the sugar is formed the glycogen should diminish; there should be a definite relation between them; this, Bernard thought he had proved.

Seegen has disproved this in the following manner:— Having first ascertained that glycogen is always evenly distributed throughout the liver, and is not accumulated in certain parts of it, he took a piece of the liver of a recently-killed or of a still-living animal, weighed it, scalded it, and determined the glycogen and sugar it contained. This operation was repeated at intervals with other portions of the liver. In this manner he discovered that the amount of sugar increases steadily from the moment of death, a process which appears to correspond with the death of the liver tissue and the cessation of its metabolic activity. On the other hand the glycogen does not undergo any corresponding diminution, except in rabbits in which it begins to dwindle at once and diminishes rapidly as the production of sugar increases.

Seegen therefore denies that glycogen is the source of sugar, and suggests that the liver converts peptone into sugar. He found that when peptone solutions were introduced into the stomachs of dogs the liver sugar was increased by from 50 to 200 per cent.

He also observed that when peptone solution was injected into the portal vein in 30-40 minutes the liver sugar increased from 100 to 300 per cent.

He has also estimated the amount of sugar which passes into the hepatic vein of a dog, and has found that the amount of carbon in the food of a dog on meat diet, neither losing nor gaining weight, is sufficient to account for the carbon of the sugar.

He now holds that sugar formation is a normal function of the liver, the quantity being considerable, 1000 grammes in twenty-four hours for a man weighing 80 kilos, the sugar being decomposed in the body.

He believes that this is a true liver sugar distinct from that derived from glycogen, that it is formed in the liver *post mortem*, but not at the expense of the glycogen present, that it does not disappear if the animal is starved or fed entirely on fatty food, and that it is formed from peptone, for after feeding with peptone its quantity becomes increased three-fold.

EXPERIMENTAL GLYCOSURIA.

Glycosuria may be produced experimentally on animals by various lesions.

Since Claude Bernard discovered that sugar appeared in the urine of an animal after puncture of the floor of the fourth ventricle, an immense mass of facts of a similar order has been accumulated, so that now it would seem as if almost any lesion of the nervous system, central or peripheral, may cause glycosuria, while a whole host of toxic substances have been shewn to possess, with more or less certainty, the same power.

In addition to the "Diabetic Puncture," the following lesions of the nervous system are stated to be followed by glycosuria :—

1.—Injury to the *vermiform process* of the *cerebellum* (Eckhard).

2.—Section of the *Spinal Cord* at various levels (Schiff).

3.—Section of the *anterior cervical nerve roots* causes permanent glycosuria; section of corresponding *posterior roots* causes only temporary glycosuria (Schiff), artificial neuritis of the *first pair of dorsal nerves* (Arthaud and Butte).

4.—Destruction of various *sympathetic ganglia, e.g.*, the superior and inferior cervical (Pavy); the first thoracic (Eckhard); the abdominal (Klebs).

5.—Section of the splanchnic nerves (Hensen), or ligaturing them (Arthaud and Butte).

Glycosuria caused by puncture of the floor of the fourth ventricle may be set aside by section of the splanchnic nerves. Other experimenters too, have found that section of the splanchnics is not followed by glycosuria, and Cyon explains this by supposing the operation to give rise to such general dilatation of the intestinal blood vessels that there is not blood enough to increase the circulation through the liver. He says if the hepatic vessels be first dilated and the spinal cord and splanchnics then cut the formation of sugar is not arrested.

6.—Glycosuria in certain cases follows irritation of the *right vagus* nerve (Arthaud and Butte).

7.—Similar results may follow section and stimulation of the central end of an ordinary sensori-motor nerve such as the sciatic (Schiff).

8.—According to von Mering and Minkowski the complete removal of the *pancreas* in dogs is always followed by glycosuria, but this is prevented by leaving a small part of the gland even though the duct be removed. The result therefore cannot be due to the absence of pancreatic juice in the intestine. Lépine suggests that the pancreas forms a sugar ferment which is absorbed by the veins and carried to the liver by the portal vein. He performed the following experiment:—Two dogs were chosen of equal size and kept fasting for 36 hours, the pancreas was then completely removed from one, and both were left unfed for 60 hours, when they were bled to death. In the blood from the one without the pancreas there was nearly three times as large a percentage of sugar as in the blood of the other, while in fifteen hours the blood of the latter had lost 33 per cent. of its sugar, while that of the former dog had lost only 6 per cent.

Minkowski has given the results of further experiments on this point. After partial extirpation of the pancreas as a rule diabetes mellitus does not appear, even when the remaining portion is no longer connected with the intestine. Still, these remaining portions must not be too small. In two cases in which the remaining pieces of the pancreas had only $\frac{1}{12}$th or $\frac{1}{5}$th the size of the normal organ glycosuria in its severest form was observed similar to that after total extirpation. It is doubtful in these cases whether the remaining portions retained their functions.

In two other cases after partial extirpation a transitory glycosuria occurred, which only continued a few hours, and later even after abundant carbo-hydrate food it did not return; this transitory glycosuria may perhaps be regarded as a result of the injury which the rest of the pancreas had suffered by the operation.

In one case, by partial extirpation of the pancreas, diabetes was produced corresponding to the slight form of this disease in man. In a dog weighing 12 kilos, a portion of the pancreas, 31 grammes in weight, was removed; only the outermost point of the tail end remaining. No sugar appeared in the urine for the next five days, even when the dog was fed with meat and milk. But after carbo-hydrates sugar appeared in the urine. After 20 grammes of grape sugar, 78 appeared in the urine. By flesh diet the sugar disappeared from the urine, to return again in great amount after a diet of cane sugar. There was therefore, in this case, after the partial extirpation of the pancreas, an injury of that specific function of this gland, the complete deficiency of which, after total extirpation, caused the severe form of glycosuria.

Lépine also removed the pancreas from a starving dog, the operation being followed by glycosuria, which was temporarily controlled by injecting into the jugular vein chyle taken from the thoracic duct of another dog during digestion. He also found that chyle added to a solution of glucose maintained at a temperature of 38° C, caused a sensible diminution of the amount of sugar in some hours. He also found that malt diastase diminished the glycosuria of an animal rendered diabetic by the removal of the pancreas. Chyle from an animal deprived of its pancreas had no effect upon sugar. Lannois, at the instigation of Lépine, made subcutaneous injections of pilocarpine on a diabetic woman with the result that there was a very notable diminution of the glycosuria, in consequence, it is suggested, of the stimulation of the pancreas by this drug.

Arthaud and Butte dispute this theory of Lépine on the ground of experiments in which they ligatured all the pancreatic veins of a dog, without increasing the amount of sugar.

On the other hand by ligaturing all the branches of the cœliac axis, except the hepatic artery, so as to cause an excessive flow of arterial blood to the liver, they caused glycosuria, and the animal died after three months. The pancreas shewed no alteration, as its circulation had been re-established by the mesenteric artery which had not been tied.

9.—Hay has found that the injection of neutral salt solution into a ligatured loop of intestine was sometimes followed by glycosuria.

10.—Glycosuria may in addition be determined by procedures directly occasioning increased flow of blood to the liver, as in the following examples:—

a.—Tying the accessory branch of the portal vein in frogs, so as to make the whole of the abdominal blood pass through the liver (Schiff).

b.—Irritation of the liver with needles (Schiff) or electricity (Pavy).

c.—Compression of the aorta or portal vein.

d.—Injecting defibrinated arterialised blood into the portal vein.

The same phenomena may also be caused by a great many toxic substances:—

a.—By the inhalation of chloroform, of carbonic acid, carbonic oxide, sulphuretted hydrogen and carbon disulphide.

b.—By strychnine (M. Foster), salicylic acid (Burton), phosphoric acid (Pavy), turpentine (Almèn,) corrosive sublimate (Rosenbach), uranium nitrate (Leconte), benzol, acetone, aldehyde, ether, chloral, amyl nitrite, amyl alcohol, morphia, opium, and curare.*

c.—Dimethyl æthal carbinol (Thierfelder).

d.—Phloridzin.

Phloridzin is a glucoside discovered in 1885, by von Kormick, in the bark of apple, pear, cherry and plum trees. It forms silky, shining, needle-like crystals, soluble in cold water. Its formula is—

$$C_{21}H_{24}O_{10} + H_2O$$

which may break up into—

$$C_{15}H_{14}O_5 \text{ and } C_6H_{12}O_6$$
Phloretin.   Phlorosin.

---

* Ashdown believes that in many cases it is glycuronic acid and not glucose which appears in the urine in these circumstances (op. cit).

The latter substance has the formula of sugar, and like it reduces copper salts and undergoes fermentation. When phloridzin is administered to animals it causes glycosuria, even when the liver has been rendered free from glycogen, or extirpated. It appears to act by preventing the processes by which sugar is stored up (as glycogen) or is burnt off in the body. The action of this substance has been studied by von Mering, who considers that he has proved that the substance excreted is really grape sugar, and not phlorosin.

Phloridzin caused glycosuria when given by the mouth or injected subcutaneously, or introduced into the veins. The sugar in the blood was not increased, while the hepatic glycogen diminished. The glycosuria continued even when the animal was starved for a long period. Occasionally in long-standing cases acetone and β-oxybutyric acid were present in the urine, and symptoms like diabetic coma occurred. The sugar was shewn to be derived from the albumens of the body.

Phloretin is the only decomposition product of phloridzin which produced glycosuria; phlorose, phloritinic acid and phloroglucin being inert.

After the administration of phloridzin and chloralhydrate, sugar and uro-chloral acid appeared in the urine.

Phloridzin was given to three persons, and sugar in considerable quantity formed in their urines. One gramme of phloridzin caused a daily excretion of 97 grammes: on the day after the phloridzin was stopped the sugar disappeared. No ill effects were produced through the use of the drug, which was continued for a month.

With reference to some of these effects it is of interest to note that Masoni states that curare diabetes may be prevented by the previous administration of arsenic, while Kirk thinks that the substances found in the urine of persons taking salicylates are chiefly other reducing bodies, but that a trace of sugar may perhaps be produced by the action of the salt on the blood corpuscles.

It has been already mentioned that ligature of the bile ducts causes glycogen to disappear from the liver, and under these circumstances puncture of the floor of the fourth ventricle fails to cause glycosuria (Wickham Legg, von Wittick, E. Külz, Frerichs). Wyatt has recently related the case of an old diabetic lady in whom the sugar disappeared from the urine during an attack of jaundice, but this, though usual, does not always follow, as the following case shews:—

CASE 1.—*Diabetes Mellitus—early diarrhœa—intercurrent jaundice—persistence of glycosuria.*

William J., aged forty-five, blacksmith, attended as an out-patient on June 12th, 1887, complaining of pain in the loins and hypochondrium. His illness began twelve months before with diarrhœa, and three months later this was followed with jaundice, which had persisted. His urine was dark amber, acid, 1026, loaded with sugar, containing a faint haze of albumen and a little bile pigment. This case shews that jaundice from obstruction is not always followed by disappearance of sugar.

THEORIES OF GLYCOSURIA.

The true bearing of all these facts upon the pathology of diabetes must remain for the present unsettled.

Let us take note for a moment of the gaps in our knowledge.

1.—It is not known definitely in what form the products of saccharine and starchy food leave the liver, whether as sugar in small quantities to be rapidly destroyed, or as some modification of glycogen.

2.—It is not known by what paths the influence of the various nerve lesions, which produce glycosuria, reaches the liver, though this is probably through the pneumogastrics (Arthaud and Butte).

3.—We are still uncertain of the nature of the influence (if any) of the pancreas on the sugar produced by digestion in the alimentary canal.

Until these questions have been settled we have not a proper basis for a rational pathology of diabetes.

Many facts favour the theory that vaso-motor paralysis of the branches of the hepatic artery is the essential feature in the production of excessive discharge of sugar from the liver, but Michael Foster points out that strychnine poisoning, in which the vessels are strongly contracted, causes glycosuria. It is possible that this may be explained by the convulsions causing a rapid discharge of sugar before the vaso-motor spasm has taken place; but the objection should be noted.

Hamilton, accepting the vaso-motor theory, says there are two possible modes in which it may act: (1) By diminishing glycogenesis, so that the sugar brought to the liver leaves it unchanged; and in support of this he quotes Ehrlich's observation that very little or no glycogen can be found in fragments of liver withdrawn by a trochar from diabetics during life; (2) That the process of converting glycogen into sugar is in excess. This latter view assumes that glycogen is normally converted into sugar, which is not certain.

Based on the same doctrine that the liver normally forms sugar are the theories of Cantani and B. W. Foster, according to whom in diabetes the liver forms an abnormal sugar, *paraglucose*, which is not oxidisable; and that of Seegen, that under certain circumstances the tissues lose their capacity for assimilating sugar.

Apart altogether from this hepatic pathology are the theories of Ziemssen and Latham, who place the production of sugar in the muscles. According to Latham a degeneration takes place in the muscle albumen due to vaso-motor paralysis and vascular dilatation, leading to imperfect oxidation and the production of sugar in the following manner: the lowest cyan-alcohol $CH_2\begin{cases} OH \\ OH \end{cases}$ instead of being oxidised to CO_2 and H_2O, is stopped at $CH_2\begin{cases} OH \\ COOH \end{cases}$, glycollic acid; this, by further oxidation forms methyl aldehyde, thus:—

$$CH_2\begin{cases} OH \\ COOH \end{cases} + O = CO_2 + \begin{cases} H \\ COH \end{cases} + H_2O$$

Then six molecules of methyl aldehyde may condense, as has been shewn to occur in plants, to form glucose thus:—

$$6 HCOH = C_6H_{12}O_6$$

These are all theories which in the present state of our knowledge can be neither accepted nor rejected.

BIBLIOGRAPHY.

ARTHAUD and BUTTE. Recherches sur la Pathogénie du Diabète, du syndrome clinique et des lésions anatomo-pathologiques déterminées chez les animaux à la suite de la névrite des nerfs vagues. "Arch. de Physiol.," 1888, p. 344.

ASHDOWN. Laboratory Reports of the Royal College of Physicians. Vol. II., Edinburgh, 1889.

BERENGER-FERAUD. Diabetes in an Ape. "Comptes Rendus," I., 1864.

BERNARD (CL.). Sur le Diabète. Leçons de Pathologie Expérimentale. Tome I. Paris, 1855.

BOUCHARDAT. Nouveau Recherche sur le Diabète. "Compt. Rendus," 1841.

BRÜCKE (E.). Vorlesungen über Physiologie. I. Bd. 2nd edition. 1875.

CANTANI (A.). Researches on Diabetes Mellitus. "Brit. Med. Jour.," 1871, I., p. 208.

DALTON (J. C.). A Treatise on Human Physiology. 7th edition. London, 1882. P. 206.

DASTRE (A.). Recherches sur les ferments hépatiques. "Arch. de Phys.," 1888, Vol. I., p. 69.

DOBSON (M.). Experiments and Observations on the Urine in Diabetes. "Med. Ob. and Enq.," Vol. V. London, 1776.

DOMINICIS (N. de). Studii sperimentali intorno agli effetti delle estirpazioni del pancreas negli animali. Diabete mellito sperimentale. " Giorn. intern. delle Sci. Med." 1889, p. 801.

ECKHARD (C.). Die Stellungen der Nerven beim künstlichen Diabetes. "Beit. zur Anat. und Phys.," Bd. IV., 1867.

FINKLER. Fifth Congress for Internal Medicine, Wiesbaden, April, 1886.

FORT (A.). Diabète consecutif à la chloroformisation pour une opération de rétrécissement de l'urètre. " Gaz. des Hôp.," 1883, No. 148.

FOSTER (B. W.). Clinical Medicine: Lectures and Essays. London, 1874.

FOSTER (M.). A Text-book of Physiology, Part II., Book 2. London, 1889.

FRERICHS (F. T.). Ueber den Diabetes. Berlin, 1884.

—————. Ueber den Plötzlichen Tod und über das Coma bei Diabetes. "Zeit. für Klin. Med.," Bd. VI., Heft 1, 1883.

GREGORY (W.). First Lines of the Practice of Physic, by William Cullen, continued and completed. Edinburgh, 1829.

GRIESINGER. Studien über Diabetes. " Arch. d. Phys. Heilk.," 1859.

HAMILTON (D. J.). A Text-book of Pathology, Vol. I. London, 1889.

HIRSCH (A.). Handbook of Geographical and Historical Pathology. London: New Syd. Soc., 1883—86.

KIRK (R.). On Artificial Glycosuria. " The Lancet," 1888, II., p. 87.

KLEBS (E.). Handbuch der path. Anatomie. 1870.

KÜLZ (E.). Beiträge zur Lehre von der Glycogenbildung in der Leber. " Arch. f. Physiol.," XX.

LANDOIS (L.). A Text-book of Human Physiology, edited by Stirling. 3rd edition. London, 1888.

LATHAM (J.). Facts and Opinions concerning Diabetes London, 1811.

LEGG (J. WICKHAM). On the Bile, Jaundice, and Bilious Diseases. London, 1880.

MASONI (E.). Diabète artificiel, moyens de le produire, et moyens qui pourraient empêcher sa production. " Revue Méd. de Louvain," 1884.

MEAD (R.). A mechanical account of Poisons. London, 1747.

PAVY (F. W.). Researches on the Nature and Treatment of Diabetes. 2nd edition. London, 1869.

ROLLO (J.). Cases of the Diabetes Mellitus. 2nd edition. London, 1798.

SCHIFF (J. M.). Untersuchungen über die Zuckerbildung. Würzburg, 1859.

SCHMIEDEBERG and MEYER. Ueber Stoffwechselprodukte nach Campherfütterung "Zeitschr. f. Physiol. Chemie," Bd. III., p. 422.

SEEGEN. Der Physiologische Auflage der Diabetes Mellitus. "Zeitschr. f. Klin. Med.," Bd. VIII., p. 328—363, and Bd. XIII., p. 267; also " Pflüger's Archiv.," Bd. XL., p. 48.

SENATOR (H.). Ziemssen's Cyclopædia of the Practice of Medicine, Vol. XII. London, 1887.

THIERFELDER (H.). Ueber die Bildung der Glycuronsäure beim Hungertier. "Zeitschr. f. Phys. Chem.," Bd. X., p. 134.

VON MERING (J.). Ueber Diabetes Mellitus. "Zeitschr. f. Klin. Med.," Bd. XIV., p. 405.

—————— and MINKOWSKI (O.). Diabetes Mellitus nach der Extirpation des Pancreas. "Arch. f. exp. Path.," Bd. XXVI., Heft 5 and 6; and "Centralb. für Klin. Med.," June 7th, 1890.

WEDENSKI. Zur Kenntniss der Kohlehydrate im normalen Harn. "Zeitschr. für Physiol. Chemie.," Bd. XIII., p. 122.

WILLIS (T.). Pharmaceutice rationalis sive diatriba de medicamentorum operationibus in Humano Corpore. Oxonio, e theatro Sheldoniano, prostant apud Ric. Davis, 1679. Part I., Sect. IV., Cap. III., p. 85.

WITTICH. Ueber das Leberferment. "Pflüger's Archiv.," Bd. VII.

WYATT (W. T.). The connection between Glycosuria and Biliary Obstruction. " The Lancet," 1886, Vol. I., p. 918.

Chapter II.
ETIOLOGY.
PREDISPOSING CAUSES.

Geographical Distribution.—Very little has been hitherto published respecting the geographical distribution of diabetes. This is in part due to the absence of trustworthy statistical data and also because the incidence of a disease, the mortality of which is on an average less than 1 per 50,000 of population, is a subject upon which the experience of individual practitioners must be a fallacious guide. By the kindness of numerous friends I have been able to obtain statistics from which the following figures are calculated.

TABLE I.

Shewing the mortality from Diabetes Mellitus in some of the principal cities of Europe.

Name.	Rate per 100,000 of population.	Name.	Rate per 100,000 of population.
Paris	9·6	Dresden	4·6
Copenhagen *	7·2	Vienna	4·2
Leipzig	6·4	Christiania	3·9
London	5.88	Naples	3·2
Berlin	5·04	Rome	1·67

* This includes 72 provincial towns also.

These figures are calculated from the averages of five years at least, and are probably fairly trustworthy as the registration in these cities is carefully carried out.

Trousseau regarded the disease as rare in Paris, and there can be no doubt that it has greatly increased of late years, as shewn in the following table compiled by Dr. Jacques Bertillon, Chef de travaux statistiques de la ville de Paris.

ETIOLOGY. 21

TABLE II.
Mortality from Diabetes Mellitus in Paris for each 100,000 of population.

1865-69	1870-71	1872-75	1876-80	1881-85	1886	1887
4	3	3	5	8	11	12

TABLE III.
Shewing the mortality from Diabetes Mellitus in certain European countries and British colonies.

Name.	Rate per 100,000 of population.	Name.	Rate per 100,000 of population.
England	5·8	Western Australia	1·9
Ireland	3·2	New Zealand	2·6
Scotland	2·1	Jamaica	1·7
Prussia	1·3	British Guiana	0·24
Norway	1·9	Montserrat	6·0
Italy	1·6	Bahamas	0·2
Gibraltar	6·8	St. Kitts	0·2
Malta	13·1	Bermuda	0·7
Cyprus	no deaths	Natal	0·7
Heligoland	,, ,,	St. Helena	0·5
New South Wales	2·7	Sierra Leone	0·16
Victoria	1·6	Ceylon *	2·4 *
Queensland	0·8	Hongkong	0·1
Tasmania	3·2	Mauritius	1·4
South Australia	1·6		

* Said to be untrustworthy.

The figures for the Colonies are based on official statistics supplied me through the Colonial Office by the courtesy of the Rt. Hon. Lord Knutsford, Secretary of State for the Colonies, and are calculated on the average of ten years. In addition to those placed in the table the following were returned as having had no deaths from diabetes registered within the period taken (the last ten years); viz., Fiji, Trinidad, St. Lucia, St. Vincent, Nevis, Antigua, Dominica, Virgin Islands, Falkland Islands, Lagos and Labuan. There was no general return from

Barbadoes, but the mortality is estimated at 1 per 100,000 of population; Cape Town has an estimated mortality of only 0·4 per 100,000; and there is no return from the Gold Coast, but the principal medical officer reports the disease to be very rare.

With respect to Ceylon, the principal colonial medical officer, Dr. W. R. Kinsey, writes in his official despatch to the Secretary of State: "The Registrar-General's figures, if forwarded without explanation, will convey a very erroneous idea of the prevalence of diabetes in Ceylon, because the registration of causes of death is wholly unreliable, as no medical certificate is required . . . . and the natives of Ceylon are peculiarly sensitive on the subject of diabetes; they consider it disgraceful to have it, they never speak of it, and conceal its existence as long as they can, for it is looked upon as a punishment for illgotten wealth, and that it is necessarily fatal and unamenable to treatment." He goes on to say that "it is most prevalent amongst the well-to-do natives." One doctor in private practice informed him that he sees in his practice an average of two fresh cases of diabetes a week, and that temporary glycosuria is very frequent. He has no hesitation in saying that it is a common disease among all races, alike among Hindus (Tamils), Mahommedans, and Cingalese. It is certainly hereditary and most common among the well-to-do class. He is thoroughly of opinion that over-indulgence in starchy foods and sugar, combined with sedentary habits and resulting corpulence, and sexual excesses, indeed, excesses of any kind, are important factors in its causation; and he thinks that congestions of the liver which are apt to occur in hot climates in those persons who indulge freely in the pleasures of the table, favour its production. He believes it to be much more common in males than in females, and between the ages of thirty-five and fifty, while it is very rare in children.

It has not been possible to obtain any official statistics respecting the mortality or incidence of diabetes in India, but there is strong testimony to its occurrence very commonly in all parts of that country. In answer to my inquiries, Surgeon-General Cornish (late of the Madras presidency) wrote to me as follows :—

"The frequency of diabetes amongst the better classes, i.e., those who do not engage in manual labour, is a matter of common notoriety amongst Indian practitioners. The causes of the disease I am not so clear about. Diet, chiefly farinaceous, oily, and saccharine, may have something to do with it. But personal habits in regard to venery and other things may also have to be considered. It should be noted also that young India is subject to much brain stimulation.

"The craving for higher education and for learning western ideas through a foreign language may unduly stimulate the mental and reasoning faculties in bodies perhaps insufficiently nourished and undeveloped. Whatever the causes are, we practically find some of the best intellects in India prematurely extinguished by diabetes.

"Over and over again it has been within my experience to find that the most promising men amongst the lawyers and university graduates have been cut off in the prime of life by this disease.

"Since I left India, I think the subject has attracted attention in Calcutta, and I daresay the *Indian Medical Gazette*, of the last five years, may give you some facts in regard to the matter. My own observations seemed to shew that the educated and learned classes (chiefly nonflesh eaters) suffered in the greatest proportion, and that it was rare to find men of these classes lasting to a robust old age. The subject is a very interesting one, and well merits searching enquiry."

Norman Chevers wrote to the same effect, referring to its frequency among rich elderly natives in Calcutta, many of whom are fat and indolent, and immoderately fond of sweetmeats. The editor of the *Indian Medical Gazette*, writing in 1871, stated that among the upper and middle classes of natives in Calcutta almost every family had lost one or more of its members from this disease.

No statistical information could be obtained from the Straits Settlements (Singapore), but the disease is said to be not uncommon. A statement was forwarded from a Chinese doctor, in which he observed that while the disease is very rare in China, in Singapore he had met with seven or eight cases, in fifteen years' practice. Dr. Graham, of Sumatra, wrote that in seven years' practice among 15,000 Chinese labourers, he had met with only one case.

The statement of the Chinese doctor above quoted, concerning the rarity of diabetes in China, has been fully borne out by all the information I have been able to obtain. Dr. Burge, of Shanghai, was exceedingly kind in making enquiries amongst his colleagues, and their unanimous testimony is that the disease is very rare in China. It is also equally rare in Japan, so far as my information extends.

The statistics from Mauritius were followed by a report of a discussion on Diabetes, which was instituted in the local medical society as a result of Lord Knutsford's circular, and this fully bore out the testimony of the figures that the disease is rare in the island. This fact, taken together with the rarity of the disease in British Guiana, is interesting, because both countries employ a large number of natives of India as coolies on sugar plantations. It seems to afford a strong confirmation of the views expressed by Indian practitioners in favour of the exemption of those classes engaged in manual labour. From Madagascar no definite information was obtainable, but Mr. S. Blackwell Fenn wrote that it is very rare.

No statistics are to be obtained from Persia, but Dr. Tholozan kindly wrote me a letter giving me the result of his long experience in that country. He believes it is less frequent than in Europe, and more common in individuals of Turkish race than among native Persians. He attributes the disease to the immoderate use of rice, sweet sherbets and sweetmeats.

Dr. Hessenauer wrote that in five years' practice he did not meet with a single case of the disease in Palestine. The official statistics of the town of Smyrna have been applied for but are not yet to hand, though they are promised. My application to official sources for statistics for Egypt was refused on the ground of insufficient clerical assistance to compile the information, but Dr. Grant Bey kindly wrote me that the disease is not uncommon and that according to his experience it prevails chiefly among Europeans, Syrians, and Turks, the Copts and Arabs being less subject to the disease, though not exempt.

In Cyprus there had been no death from diabetes for ten years, and the chief medical officer in ten years' practice had seen only two cases, one in a Cypriot.

Respecting America, it was found very difficult to get any information of value. Tyson estimates the mortality from diabetes in Philadelphia at 11 per 10,000 deaths, and in New York 7, whereas in London the proportion is 26, and in all England 25.

According to the editor of the *New York Medical Record* the mortality from diabetes in the United States was 837 or 2·1 per 100,000 living in 1870, and 1443 or 2·8 per 100,000 living in 1880. The article goes on to say that glycosuria is very prevalent in the north-west where the air is very dry, business competition exceedingly keen and nervous wear and tear consequently very great. It is to be feared that at present no statistics can be obtained in the United States which can be fairly compared with the figures obtained from European sources. The disease seems to be rare in California, the statistics of San Francisco returning the mortality at only 0·6 per 100,000 living. Dr. Gresham, of Los Angeles, wrote that in seven years' practice he had not met with a single case. From South America it is even more difficult to get any information. Dr. Louis Colbourne, of Buenos Ayres, wrote that there are no statistics, but that in his opinion it is very rare, and Dr. Harvey reported the same of Chili. Its rarity in British Guiana is shewn by the figure in Table III., and this is confirmed by a private letter from Mr. C. Macnamara, lately Colonial Surgeon at Essequibo.

This information, by no means complete, has been obtained with a vast amount of labour, and probably represents all or nearly all that can be learnt upon the subject at present.

Can any deductions be safely drawn from these facts? It is very doubtful. Certainly there seems some evidence that the Mongolian and African races enjoy a certain degree of immunity, but this must wait confirmation.

The high rate of mortality in Malta is very remarkable, and it is curious that the report was not accompanied by any comment by the officer who reported the figures.

There might at first sight appear to be a decided advantage in the country over the towns, as the great cities have a far higher average than the countries in which they are situated; but it must be remembered that the country statistics are always less reliable than the towns, and if we look to English figures we find reason to think that there is no such immunity in country districts. Taking the average of England and Wales as 25 per 10,000 deaths, we find the following counties above the average:—

Berkshire	... 100	Lincolnshire	...	70
Cambridgeshire	... 90	Buckinghamshire	...	64
Oxfordshire	... 80	Suffolk	...	60
*Rutland	... 80	Sussex	...	60

*The figures are too small to be of value.

ETIOLOGY. 29

TABLE VI.

Age.	Actual Number.			Percentage Ratio.		
	Males.	Females.	Total.	Males.	Females.	Total.
Under 10	3	5	8	0·22	0·36	0·58
10 to 20	35	22	57	2·57	1·61	4·19
20 ,, 30	69	28	97	5·07	2·05	7·18
30 ,, 40	154	70	224	11·32	5·14	16·47
40 ,, 50	260	79	339	19·11	5·80	24·92
50 ,, 60	281	137	418	20·66	10·07	30·73
60 ,, 70	138	44	182	10·14	3·23	13·37
70 ,, 80	25	9	34	1·83	0·66	2·49
Over 80	1	—	1	0·07	—	0·07

The following table is taken from the Report of the Registrar-General for England for the year 1886.

TABLE VII.

All Ages.	Under 1	1	2	3	4	5	10	15	20	25	35	45	55	65	75	85 and upwards.
Males 978	2	2	5	1	—	8	22	40	43	110	118	143	246	189	49	—
Females 656	—	—	2	—	1	9	26	41	38	62	91	101	136	101	44	4
Total 1634	2	2	7	1	1	17	48	81	81	172	209	244	382	290	93	4

These tables all agree very closely, though the maximum mortality, as might be expected, occurs a few years later than the maximum of living cases. Thus Schmitz and Pavy agree in finding the largest number of cases between the ages of forty and sixty, while the maximum mortality is from fifty-five to seventy-five. This would appear to indicate that such elderly diabetics live a considerable time, which is strictly in accordance with our clinical knowledge.

Sex.—Diabetes is more common among men than women, in the proportion of rather more than 3 to 2 But this disparity does not seem to hold good amongst children. In Table VII. the numbers under twenty give 80 males to 79 females, the proportion under fifteen being 40 males to 38 females. In Table VI. (Pavy's) there were 3 males to 5 females under ten years of age, though under twenty there were 38 males to 27 females; in Stern's 117 cases the girls were more numerous than the boys in the proportion of 5 to 3 ; as age advances the proportion becomes nearly 2 males to 1 female, until in old age it comes back to the equality of childhood.

Heredity.—There can be no doubt that diabetes mellitus is hereditary. Chevers quotes the following statement from the *Indian Medical Gazette* for 1871 : " In one of the most influential native families in Calcutta, No. 1 died of acute diabetes at the age of sixty-five ; No. 2, his eldest son, died of chronic diabetes ; No. 3, his third son, had chronic diabetes ; No. 4, his fourth son, died of acute diabetes ; No. 5, his daughter, aged sixty-five, had diabetes ; No. 6, eldest son of No. 2, died of chronic diabetes ; No. 7, second son of No. 2, had diabetes ; No. 8, eldest son of No. 4, had diabetes." Thus the disease affected eight members of the same family, extending over three generations.

Auerbach has related the history of a family, in which the father died insane and the mother diabetic. Of eleven children, one died insane, one of diabetic coma, four others were insane, and one was diabetic. Frew has recorded a more remarkable case in a little girl aged nine, suffering from diabetes, whose paternal grandfather, uncle and aunt, had all died of the same disease.

Nevertheless in the majority of cases there is no evidence that the disease has been inherited, and it is more common to obtain a history of a diabetic brother or sister than of a diabetic parent. This is perhaps due to obvious causes, as it is always exceedingly difficult to get information which goes back any number of years.

Trousseau thought that phthisical parents often had diabetic children, and there can be no doubt that a history of tubercular disease can often be traced in diabetics.

Excessive use of Sugar and Starchy Food.—There is a very widespread belief that the excessive use of sugar tends to diabetes. Eichhorst has met with the disease among the workpeople in sugar factories. Christie, who first described the prevalence of the disease in Ceylon, referred it to the excessive use of sugar and starchy food. Tholozan regards the disease among the Persians as caused by sweetmeats and sweet sherbets. Indian practitioners recognise the effects of the same causes; yet, as has already been mentioned, in Mauritius and British Guiana, where the people are chiefly employed in sugar factories, the disease is very rare. Charcot states that glycosuria is frequently produced among the novices in the Monastery of La Trappe.

Beer.—Kratschmer thinks that the urine of great beer drinkers always contains more or less sugar, but there is certainly no evidence that beer drinking *per se* is a common cause of diabetes. Were it so, it is scarcely conceivable that the kingdom of Prussia would occupy such a relatively favourable position in the table.

Cider has been said to have some relation to diabetes, and the prevalence of the disease in Normandy is ascribed to this cause, but the cider counties in England shew a mortality much below the average.

Obesity.—Many diabetics are, or have been, very stout. It is generally in elderly persons that this connection is observed. Kisch has observed temporary glycosuria as a frequent occurrence in general obesity. Seegen says that 30 per cent. of the cases he has analysed were excessively fat at the beginning of their illness. Kisch says that in all cases where hereditary obesity comes on in youth and progresses rapidly, and in non-hereditary cases where treatment is not beneficial, diabetes must be looked for as a possible occurrence. He believes one-half of the former and 15 per cent. of the latter become diabetic. In many families it can be shewn that some members are fat from childhood and others are diabetic, or the fat individuals become diabetic about the fourth decade. He inclines to the belief that the muscles becoming infiltrated with fat, are unable to convert sugar into glycogen.

EXCITING CAUSES.

Frerichs gives the following classification of the causes of glycosuria.

1. *Toxic Glycosuria.*
>Curare.
>Carbonic oxide.
>Amyl nitrite.
>Ortho-nitro-phenyl-proprionic acid.
>Methyl-delphinin.
>Morphia.
>Chloral.
>Hydrocyanic acid.
>Sulphuric acid.
>Mercury.
>Alcohol.

Also from morbid poisons.
>Cholera.
>Anthrax.
>Diphtheria.
>Enteric fever.
>Scarlatina.
>Malaria.

2. *Glycosuria from Digestive disturbance.*
>Gastric catarrh.
>Gout.

3. *Glycosuria from Nervous disturbance.*
        Psychoses.
        Neuralgias.
        Injuries to brain and cord.
        Brain diseases. Apoplexies, aneurisms of cerebral arteries, meningitis cerebro-spinalis, etc.

With reference to the list of toxic glycosurias, it is possible that glycuronic acid has in some instances been mistaken for glucose.

Chloroform.—This may not only give rise to temporary glycosuria, but occasionally causes acute and fatal diabetes (Fort).

Bromide of Potassium.—Weber has described a case of acute and fatal diabetes in a little girl of seven, the effect of swallowing an ounce of this drug, which she took by mistake.

Opium.—Campbell mentions a case in which the administration of opium was always followed by glycosuria.

Contagion.—It is very contrary to what little we know of the pathology of diabetes to suggest that it is a disease which can be transmitted by contact, yet Debove has drawn attention to the fact that out of fifty cases of diabetes, in five the husband and wife were attacked simultaneously. Lecorché, who had observed the same facts, attributed them to community of anxieties and nutriment; but Debove thinks that the lives of men and women are very different, and he remarks that husband and wife are rarely both gouty; though in England this is not uncommon, and possibly husbands and wives live a life more in common in England than in France. However, these observations on the frequent occurrence of diabetes in husband and wife are supported by Dreyfous, Gaucher, Labbé, Letulle, Schmitz, and Rendu, and the inference is that under certain circumstances, or in some obscure way, diabetes may be contagious. Such a view is exceedingly improbable, and the facts do not warrant our adopting it. It is well-known that husband and wife become insane with a degree of frequency that is very remarkable, but even our lively Gallic neighbours do not, in that case, suggest that insanity is contagious. They are content to speculate about a "psychical affinity," which apparently draws together those predisposed to insanity. The truth is that diabetes is getting to be a common disease in certain classes, especially the wealthier commercial classes, so that such coincidences are likely to be met with.

Injuries.—Severe contusions, especially when due to falls, have often been followed by diabetes. Pavy has recorded a case following a kick from a horse. In a case of Ebstein's, a boy aged fourteen was struck by a stone in the region of the stomach. This was followed by pain and increase of the liver dulness, which gradually disappeared; but three months later diabetic symptoms set in. A patient of Lindsay's fell and struck his right loin. This was followed by pain and local swelling, which disappeared in time, to be in turn followed by thirst, polyuria, and emaciation. In some cases there is reason to believe that the nervous centres may have been injured, as in a case of Dale's, where a child fell on the back of its head, and months later acute and fatal diabetes set in; and in the case of the officer related by Moosdorf, who became diabetic after a bullet wound in the neck. Scheuplein has published the case of a young dragoon who fell from a window forty feet high, and dislocated the twelfth dorsal vertebra. He was treated by extension, but eleven days after the injury acute diabetes set in, which lasted about a month, and then gradually disappeared, the patient making a good recovery from his accident, and being quite well two years afterwards.

CASE 2.—*Diabetes Mellitus—due to a blow—thirst—polyuria—wasting.*

John B., fifty-four, wire drawer, admitted as an out-patient on March 12th, 1889, complaining of thirst and polyuria. He had been ill six months, and had been treated with brown bread *ad libitum* and codein (gr. i.) three times a day. Just before this illness began he bruised his hip at his work. There was no family history of diabetes, but one of his grandfathers had gout. The quantity of water varied from 4½ to 6 pints, sp. gr. 1032-35, loaded with sugar. He weighed 7st. On probably the same diet, though he was told to eat bread very sparingly, Jambul perles (Christy) given in as large a dose as four, three times a day, produced no effect ; and as much could be said of antipyrin. The quantity of water gradually diminished to three or four pints, and his weight was 6 st. 13 lbs. on April 30th, 1889.

According to Redard, ephemeral glycosuria is of frequent occurrence in surgical practice, even after slight injuries and subcutaneous fractures, but especially in connection with phlegmonous and gangrenous affections, anthrax, lymphangeitis, and erysipelas.

CASE 3.—*Diabetes Mellitus—following erysipelas—loss of sexual desire—slight polyuria—slight thirst—no knee jerks—no wasting.*

Mr. Paul R., aged forty-eight, consulted me in March, 1888, complaining of diabetes, of which he had been aware for three or four months after an attack of erysipelas supervening in his foot from a cut got by treading on broken glass. The polyuria and thirst were slight, and there was no wasting ; his chief complaint was his loss of sexual desire, which seemed to depress him mentally, although he was unmarried. His knee jerks were absent. The urine measured from four to five pints daily, sp. gr. 1029, loaded with sugar.

Redard thinks there is a decided relation between pus formation and glycosuria. He has recorded a case in which a small abscess by the side of the rectum was accompanied by polyuria (2½ litres) and glycosuria, which symptoms disappeared when the abscess had completely healed. But in certain instances true diabetes may follow, as in a case recorded by Frerichs, where the exciting cause was an abscess connected with a tooth. Spencer says passing glycosuria is common in inflammatory affections of the skin and subcutaneous tissue. It also appears towards the end of cases of septicæmia, but is most frequent in association with carbuncles and severe boils, in patients whose urine was previously free from sugar. M'Nish has recorded the case of a blacksmith, aged fifty-six, who was severely burnt by a bottle of paraffin oil taking fire. Twelve days after the accident his urine became saccharine, and this continued for twenty-five days, until the burn was healed. Albert and de Renzi have observed glycosuria occurring in gangrene, and Spencer has noted it in connection with malignant disease not affecting internal organs, where it disappeared after local treatment. The following case is an example of glycosuria in connection with malignant disease :—

CASE 4.—*Malignant Disease of Abdomen—glycosuria—no polyuria—slight thirst.*

William J., forty-five, admitted on July 12th, 1887, complaining of having had diarrhœa for a year, and of passing slime and blood from his bowel.

History.—He had lost five stone in weight, and had severe pain in the loins and hypogastrium, so great as to make him cry out. There had been no swelling, but nine months before he had jaundice, which lasted two or three months; he used to vomit, but he had no pain then. Two years ago he was laid up with bronchitis, but could remember no other illness. Of late he had suffered from thirst, but did not think the quantity of his water excessive. Father dead, aged fifty, cause unknown. Mother living, in good health. One sister died in the Queen's Hospital from some wasting disease. He had never heard of any case of cancer, gout, diabetes, fits, insanity, or consumption in his family.

Condition on Admission.—He was a stoutly built, fairly nourished man, weighing 9 st. 8 lbs.; complexion sallow; conjunctivæ pearly; no œdema; skin pale; no pigmentation. Pulse 84, Resp. 18, Temp. 98°. Tongue furred posteriorly; no pain or discomfort after food; appetite not very good; bowels loose, open five times since admission; stools brown, fluid, containing a little blood. Liver dulness only 2½ inches in V.M.L.; some dulness in flanks, which disappeared on change of position; spleen not enlarged. An oval swelling could be felt above Poupart's ligament on the left side; it was badly defined towards the middle line, and apparently passed down into the pelvis; it was tender, and the seat of pain. Heart's apex in the fifth left intercostal space; no murmur. Lungs resonant; breath sounds normal. Urine, 54 oz., sp. gr. 1035, acid, orange coloured, white deposit, consisting of leucocytes and squamous epithelium and granular matter; urea, 1·5 per cent.; a cloud of albumen; sugar, 5 per cent. He was dieted, and given chalk and opium mixture. On this treatment the sugar disappeared from his urine.

Progress of Case.—On the 27th he was put on ordinary diet, when sugar returned at once; his diarrhœa became worse, and his temperature began to be irregular.

On July 31st he was ordered to resume strict antidiabetic diet.

Aug. 4th. He had a rigor, and his temperature at night was 101·2°. After this his evening temperature was generally over 101°, the sugar was always present, but the quantity of urine remained low.

August 12th. Another tumour was felt to the left of the umbilicus. The liver had become enlarged, coming down nearly to the umbilicus, where its hard, firm edge could be felt.

Aug. 13th. We discovered that he was throwing away

his urine, and he had evidently been evading his diet restrictions for some time.

Aug. 15th. His urine contained only 1 per cent. of sugar.

Aug. 20th. He was made an out-patient, but he only attended once.

Muscular Strain.—The following case illustrates the possibility of diabetes following a simple muscular strain.

CASE 5.—*Diabetes Mellitus—due to muscular strain.*

William C. S., aged twenty-nine, bricklayer's labourer, was admitted into hospital, October 23rd, 1889, complaining of polyuria, thirst, hunger and wasting. His illness had lasted about two months.

Family History. — Father died of sunstroke in Australia, aged fifty. Mother living, aged forty-nine, in good health. There were fourteen children of whom ten died in infancy; two brothers and a sister were living in Australia; the last had been lately in hospital for some internal complaint. No history of diabetes, consumption, gout, epilepsy, insanity or cancer in the family.

Previous History.—Previous health good. Had been in the army, and was in India for five years. He had an attack of colic in 1880. Denied having had syphilis. Could remember no injury except that two months ago just before this illness began, he was carrying a scaffold pole and slipped; in trying to recover himself he twisted his back and must have bruised his testicle as it swelled afterwards, though he felt no pain.

History of Present Illness.—Patient returned from India in 1885; he married in 1886. He had worked as a bricklayer's labourer; he had had a comfortable home, but no children. After the accident referred to above, he got quite well, but shortly after he noticed that he was passing more water, and that he was getting thinner and weaker. His appetite increased, and a fortnight before admission, he began to be tormented with thirst. He was then passing ten or eleven pints of water daily, and his weight had fallen from eleven stone seven pounds to nine stone ten pounds.

Present Condition.—A well-developed, sparely nourished man, of fair complexion; skin dry; no subcutaneous fat; no œdema. Weight nine stone ten pounds. Temp. 98°. Pulse 72. Resp. 18.

Alimentary System.—Lips pale, tongue clean at tip, but covered over dorsum with a greyish brown fur; teeth very even and good, but one or two molars carious; appetite moderately good; thirst excessive; no pain or discomfort after food; bowels regular. Liver dulness in V.M.L., 3½ inch. Splenic dulness in M.A.L., 1 inch.

Circulatory System.—Apex beat hardly to be felt in 6th I.S., half-an-inch internal to the V.M.L. Heart's limits otherwise not increased. Pulse, regular, soft, feeble. Heart sounds normal.

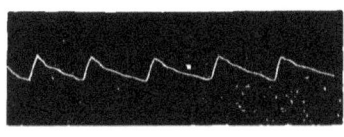
Fig. 1.

Respiratory System.—No cough; chest well formed; percussion note and breath sounds normal.

Special Senses.—Eyesight good. Deaf with both ears, but this had been coming on for four or five years, and his mother had been deaf as long as he could remember.

Nervous System.—No pain in head or spine; no giddiness. Complained of some noise in head, but chiefly in ears (tinnitus). Sleep broken, but this not altogether due to frequent micturition. No abnormal sensations; no paralysis or loss of sexual power.

Muscular System.—There was some general diminution in bulk of the muscles.

Urinary System.—He had to rise two or three times at

night to make water. Urine 180 ozs.; sp. gr. 1034; acid; very pale straw colour; no deposit; urea, 0·7 per cent.; no albumen; sugar 8·69 per cent.

He was in hospital eleven weeks, and on his discharge had gained sixteen pounds in weight; the quantity of urine was under one hundred ounces daily, and the percentage of sugar was 4·5. He was discharged on January 10th, 1890.

He was re-admitted on March 18th, 1890, having been an out-patient in the interval. On March 11th, he complained of increasing weakness and loss of flesh, and was urged to come in again. He refused to do so, but on the 13th he took to his bed, and remained there till brought to the hospital. On admission he complained of pain in the left iliac fossa, with tenderness on pressure. Tongue dry. Pulse 110. Resp. 20. Temp. 97·6. The quantity of water had gone up to about a daily average of 160 ounces, containing 5 to 6 per cent. of sugar. This pain in the left iliac fossa and groin continued till the 24th, when he was able to get up for a little.

He seemed deafer than formerly, and the ears were carefully examined.

Right ear, hearing *nil.*; this had been always the case since an abscess in India, eight or nine years ago. Membrane drawn inwards, and looked thickened.

Left ear, hearing for conversational purposes defective, and notably variable; he read a good many words on the lips, so that it depended somewhat on the people who addressed him. Heard watch at one inch from ear. Membrane slightly thickened and drawn inwards.

Weight on admission 9 st. 10 lbs.

He was again placed on animal diet with opium and belladonna. He gained weight rapidly, but the quantity of urine did not diminish, or the percentage of sugar. Acetone was constantly present. On May 4th, he stages of a case of tumour of the cervical portion of the cord.

There is a considerable amount of evidence that mental emotion may often be the starting point of an attack of diabetes. Losses in business, family affliction, danger to life, etc., have been recorded as immediately preceding the onset of the disease.

The association between hysteria and diabetes has been insisted upon by Grenier, but without sufficient evidence that this amounts to anything more than a casual coincidence. It may be, however, that the diabetic dyscrasia, like plumbism, favours the development of hysteria.

Manby has recorded three interesting cases, illustrating the connection between exophthalmic goitre and diabetes. In the first a diabetic father had two children, of whom one was diabetic and the other the subject of goitre, exophthalmos, and irregular action of the heart; in the second, one brother was diabetic, the other brother lost two children with acute diabetes, and the sister had Graves' disease; in the third, two sisters were diabetic, and a third the subject of Graves' disease. Since Hale White has shewn that there is reason to regard the floor of the fourth ventricle as the seat of the central lesion in Graves' disease, this connection is specially interesting.

Acute Rheumatism.—Diabetes may follow acute rheumatism (Hughes Bennett).

Gout.—Glycosuria is a common phenomenon in gout; it is not uncommonly attended by polyuria and thirst, but there is very slight loss of weight.

CASE 6.—*Diabetes Mellitus—polyuria—thirst—albuminuria—pruritus ani—lichen on chest and back—disappearance of sugar.*

Mr. Frederick W. W., aged fifty-six, consulted me first on March 31st, 1883, being sent to me by my friend Mr. Eales, who had discovered sugar in his urine. He had been passing too much water for five months, and suffering from thirst. He had had gout several times. Weighed 13 st. 2 lbs.; had not lost flesh. Tongue clean; appetite good; bowels regular. Fundus of eyes normal. Urine, 1028, contained sugar and a trace of albumen. On antidiabetic diet, using toast moderately, and extract of opium (gr. i.) three times a day, his urine was diminished by two-thirds, its specific gravity came down to 1020, but there was some loss of flesh—by May 5th he had lost 5 lbs. By June 16th the sugar had disappeared, and did not return at all after June 30th. On his last visit, Aug. 18th, 1884, he weighed 13 st. 3 lbs., his urine amounted to two quarts daily; it was free from sugar, and he felt very well.

Liver Diseases.—Roger asserts that he has met with glycosuria in various affections of the liver, including cirrhosis and persistent jaundice. Hull has recorded a case of severe and fatal diabetes which set in ten months after the bursting of an abscess of the gall bladder. Some years ago I attended with my friend, Dr. R. Norris, of Birchfields, a gentleman who had been suffering from diabetes, when he was suddenly attacked by jaundice, with complete obstruction of the common duct. The sugar disappeared from his urine and did not return after the bile regained its passage into the intestine.

Acute Tonsillitis.—Rogers has published a case of acute tonsillitis in a gentleman aged forty-four, which was followed by diabetes, terminating fatally from coma twenty hours after being first recognised.

Diphtheria.—Glycosuria may follow diphtheria, as in a case reported by Zinn, which, however, ended in complete recovery after lasting ten weeks.

Influenza.—I have already published the notes of two cases of diabetes following attacks of influenza in the epidemic of 1889-90. They are of sufficient interest to be reproduced here:—

CASE 7.—The earlier in point of date was that of a

girl named O. B., aged twenty-two; she had never been very strong, and at Christmas, 1889, she had an attack of influenza. Shortly after this she began to suffer from thirst and frequency of micturition, when she was passing about 100 oz. of urine daily containing 10 per cent. of sugar. The case improved under treatment but did not recover, and the condition has persisted.

CASE 8.—C. A., a glassblower, aged thirty, was in bed for a week at the end of January, 1890, with a severe attack of influenza. Immediately after this he began to suffer from thirst and loss of weight. He had been in the habit of weighing himself at intervals and was sure that up to the time of the attack of influenza he had not been getting thinner, but since then, though a spare man, he had lost 17 lbs. He was passing 200 to 300 ozs. of urine daily, containing 7 to 8 per cent. of sugar.

Malaria.—The importance of malaria as a cause of diabetes is doubtful. Burdel believes that glycosuria occurs very commonly in connection with this infection, and gives the following table of cases:—

	Cases.	Sugar.	Percentage.
Febris quotid.	134	29	22
,, tert.	122	17	14
,, quart.	76	11	14
,, perniciosa	11	3	27
Severe cachexia	40	32	80

Calmetti supports this view, stating that while acute malarial affections are sometimes attended by transitory glycosuria, severe malarial cachexia is a distinct cause of diabetes.

On the other hand, in the course of a recent discussion at the Société Médicale, of Mauritius, the great majority of the speakers were opposed to the belief in any relation between these diseases, and this seems to be the general view of British practitioners residing in malarious countries. It is strange if this relationship is true that the disease should be so rare on the West Coast of Africa and in British Guiana, while we should expect it to be more common in Italy than statistics shew it to be. It is a point upon which additional information is wanted, and it is worth noting that the malarial counties of England are those which stand among the highest in the mortality from diabetes.

Enteric Fever.—The following case of diabetes followed enteric fever.

CASE 9.—*Diabetes Melllitus—following enteric fever—œdema of legs.*

Annie S. —, aged thirty-nine, housewife, was admitted to hospital on September 23rd, 1890, complaining of great thirst, increasing weakness, pain in eyes, and dimness of vision and polyuria. Duration thirteen months.

Family History.—Father and mother both dead, the former from apoplexy, aged forty-seven, the latter from cancer of liver, aged forty-nine. Patient was one of a family of thirteen, of whom six died in infancy, some from croup, others from measles, and during teething; the remaining seven were alive and in fairly good health, but not robust. Patient was married and had had nine children, four of whom died in infancy, and five were living in good health. Her husband also was well and strong.

Previous Health.—Patient had measles when seven years old, scarlet fever when a baby, small-pox since marriage (at twenty-one), from which she quite recovered, and was in good health until thirteen months ago. In August 1889, she had enteric fever for four months, and ever since she had suffered from thirst and polyuria.

Present Condition.—Patient was a fairly well-developed woman, but ill-nourished and thin; there was no cyanosis or jaundice, but some slight œdema of legs; skin moist. Temp. 98·4. Resp. 16. Pulse 84.

Teeth good, gums red, tongue thick, marked at edges

ETIOLOGY. 47

by the teeth and covered with a thick whitish fur; appetite very good. Thirst very great; had pains in the stomach when fasting; bowels confined; abdomen full, with some tenderness over the epigastrium; there was no dulness on percussion.

Liver dulness in V.M.L., 4 inches.

Splenic dulness in M.A.L., 2 inches.

Respiration, 16. Movements of thorax, regular and equal. Percussion note resonant, V.R. and V.F. equal on both sides. Breath sounds normal; no adventitious sounds were heard, but she had slight cough.

Heart sounds clear, no murmurs were heard; cardiac dulness, upper limit, 3 C.C. Left limit internal to V.M.L. Right limit ¼ inch to right of sternum. Apex beat in 5th space internal to V.M.L.

Pulse 84, regular, full; the tracing (*Fig.* 2) shews a very fair amount of tension. She had some dyspnœa on exertion, but not to any marked extent.

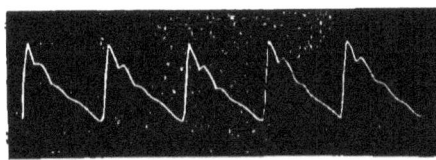
Fig. 2.

Urine quantity 76 oz.; sp. gr. 1037; acid; no albumen; loaded with sugar; complained of a burning sensation on passing water. Had always been regular until now, last time it was very scanty, and she had missed this time.

She was sensitive to pain, heat and cold; no numbness or tingling, pupils equal, slightly dilated, responded readily to light. Patient could not see to read even with her glasses, but could see and distinguish individuals in a good light and could see much better with glasses; her sight was much worse the last six weeks. Hearing very good. Taste and smell unimpaired. Patellar reflexes also absent on both sides, both plantar reflexes present. Did not sleep well.

Syphilis.—From what has been said in connection with diseases of the nervous system, it will be readily understood that gummata in the brain may give rise to glycosuria or to all the symptoms of diabetes, but Ord believes he has met with several cases of glycosuria without any evidence that the medulla was implicated. Decker has reported a case of diabetes occurring in the course of a syphilitic affection of the eye, which was cured by mercurial inunction without any dietetic restrictions.

Soon after Christmas, 1889, I saw a man who had diabetes and a syphilitic sore mouth, but there were complicating factors in the etiology, as he had had influenza, and had received a severe blow on the loins from a heavy iron bar. Moreover anti-syphilitic treatment which cured his mouth, did not cure the diabetes.

Decker attributed his case to syphilitic disease of the cerebral arteries.

Pregnancy.—The fact that the urine of pregnant and nursing women contains sugar has been long known. Bennewitz, in 1828, recorded the case of a stout young woman, in whom diabetes with severe thirst and polyuria came on during the fourth pregnancy, left her as soon as she was delivered, and recurred during the fifth and sixth pregnancies—leaving her during the intervals. Poulet has published the case of a girl of sixteen, who became pregnant at fifteen, and aborted at the eight month. Soon after she began to suffer from thirst, polyuria and emaciation. Matthews Duncan, after reviewing the whole subject, formulated the following conclusions:—

1. Diabetes may come on during pregnancy.

2. Diabetes may occur only during pregnancy, being absent at other times.

3. It may cease with the termination of the pregnancy returning sometime afterwards.

4. Diabetes may come on soon after parturition.

5. Diabetes may not return in a pregnancy occurring after its cure.

6. Pregnancy may occur in diabetes.

7. Pregnancy and parturition may apparently be unaffected in their healthy progress by diabetes.

8. Pregnancy is very liable to be interrupted in its course in diabetes, and probably always by the death of the fœtus.

There is no doubt that pregnancy may give rise to true diabetes, and that Griswold is wrong in asserting that it is always a temporary glycosuria unworthy to be called diabetes, being attended by neither polyuria nor thirst nor any other symptom of diabetes.

Sinclair says the glycosuria of lactation is always present in nursing women when the breasts become engorged with milk from weaning, death of the infant or other cause, and he quotes numerous cases in support of his statement, but Davenport says that it is only when lactation has existed for five or six months that its sudden suspension gives rise to the presence in the urine of *lactose* in large quantities.

Climacteric Diabetes.—It was supposed by Lecorché that the climacteric period favoured the occurrence of diabetes in women. Unquestionably elderly women get diabetes, but in some instances, as in the following, there is an apparent connection between the two events.

CASE 10.—*Diabetes Mellitus—glycosuria—polyuria—wasting—sudden onset at the menopause—cystic goitre.*

Elizabeth K., fifty, attended as an out-patient on May 7th, 1889, complaining of thirst, hunger, and " the flesh all leaving her." She had been ill a year and nine months. She had never had gout, rheumatism, or any serious illness or accident. She was married and had had eight children. She knew of no member of her family who had suffered from diabetes or gout. Menstruation ceased quite suddenly at the time that these symptoms came on. Patient was a fairly developed woman, with a flushed face and dry skin; much emaciated, weighing only 7 stone. There were no symptoms of dyspepsia; her bowels were regular. She had a large cystic goitre, which she had had since puberty. Her pulse was quick, 144; physical signs normal. Urine, 1035, loaded with sugar. She had to rise three or four times each night to pass water.

Lawson Tait believes that it is of essentially good prognosis, and can be treated by opium alone without dietetic restrictions.

Imlach has recorded a remarkable case of a diabetic woman, aged thirty-one, on whom he operated for pyosalpinx by removal of the appendages, and in whom the operation was not only perfectly successful in curing the local trouble, but was followed by the complete disappearance of the diabetes.

Excessive sexual indulgence is apparently, at times, an exciting cause of diabetes. Dickinson mentions the case of a young man who died at the early age of twenty-five, in whom he could find no other cause but excessive promiscuous sexual indulgence, to which he had been addicted since the age of seventeen. *Masturbation* is another alleged cause, but Yarrow has recorded a case of temporary glycosuria attributed to enforced sexual *continence*. Under treatment with bromide of ammonium, fluid extract of jaborandi, and external local applications to the neck and loins, he recovered in eight days, and on the disappearance of the sugar a spontaneous pollution occurred!

BIBLIOGRAPHY.

ALBERT (E.). Diabetes Mellitus and Senile Gangrene. "Allgem. Wiener Med. Zeit.," 1885, Nos. 1, 2, 4, 5, 8, and 9.

AUERBACH (L.). Ueber das Verhältniss des Diabetes Mellitus zu Affectionen des Nervensystems. "Arch. f. klin. Med.," Bd. XLI., 1887.

BENNETT (J. HUGHES). Clinical Lectures on the Principles and Practice of Medicine. 3rd edition, p. 898.

BENNEWITZ. Case of Diabetes recurring during successive Pregnancies. "Edin. Med. Jour.," Vol. XXX., 1828, p. 217.

ETIOLOGY. 51

BERTILLON (J.). De la fréquence des principales maladies à Paris pendant la période 1865-87. Paris. 1889.

BURDEL (E.). De l'impaludisme dans ses rapports avec la glycosurie et les traumatismes. "L'Union Méd.," 1882.

CALMETTE (E.). Des rapports entre la glycosurie, le diabète, l'oxalurie et les différentes formes de l'impaludisme. "Gaz. Hebdom.," Vol. XIX., 1882.

CAMPBELL (N. J.). Diabetes Mellitus. "Guy's Hosp. Rep.," Vol. XLIV., 1888.

CHARCOT (J. M.). The causation of Diabetes. "Jour. de Méd.," April, 1883.

CHEVERS (NORMAN). Practical Notes on the ordinary Diseases of India. "Med. Times," 1884, II., p. 741.

CHRISTIE (T.). Notes on Diabetes Mellitus as it occurs in Ceylon. "Edin. Med. and Surg. Jour.," 1811.

DALE (J. Y.). Diabetes Mellitus in an Infant. "Philad. Med. News," Dec. 10th, 1887.

DASTRE. *Op. cit.*

DAVENPORT (C. J.). Resorption Diabetes of Lactation. "St. Barth. Hosp. Rep.," Vol. XXIV., 1888, p. 175.

DEANE (W. H.). Diabetes Mellitus in an Infant. "New York Med. Rec.," Vol. XXXIII., 1888, p. 97.

DEBOVE. The Contagiousness of Diabetes. "Bull. Soc. Méd. des Hôp. Jour. de Méd.," Sept. 29th, 1889.

DECKER (C.). Zur Casuistik des Diabetes mellitus syphiliticus. "Deut. Med. Woch.," 1889.

DICKINSON (W. H.). Diabetes. London. 1877.

DUNCAN (J. MATTHEWS). On Puerperal Diabetes. "Obstet. Trans.," Vol. XXIV., 1882, p. 256.

EBSTEIN (W.). Ueber Drüsen-epithelnekrosen beim Diabetes mellitus, &c. "Deut. Arch. f. klin. Med.," Bd. XXVIII., 1881.

EDWARDS. Glycosuria in the course of Insular Sclerosis. "Revue de Méd.," 1886, p. 935. (See Roger.)

EICHHORST. Praktische Erfahrungen über die zuckerige und einfache Harnruhr. "Corresp.-Blatt f. Schweizer Aertzte," 1888, No. 13.

FAGGE (C. H.). The Principles and Practice of Medicine. 1886.

FERRARO (PASQUALE). Il Diabete Mellito spontaneo e mortale degli animali paragonato con quello dell' uomo. "Il Morgagni," February and April, 1885.

FORT (A.). *Op. cit.*

FRERICHS (F. T.). *Op. cit.*

FREW (W.). Case of Diabetes in a girl of nine with hereditary tendency. "Glas. Med. Jour.," April, 1887.

GRENIER (R.). Hystérie et Diabète. "Arch. Gén. de Méd.," 1888, II.

GRISWOLD (G.). Diabetes in Pregnancy. "New York Med. Jour.," 1884, II., p. 623.

HAMILTON (A. M.). The significance of Glycosuria in connection with Disease of the Brain and Cervical Cord, and with Dementia Paralytica. "New York Med. Jour.," 1884, II., p. 1.

HERMANIDES (S. R.). Eene ziekte-geschiedenis waaruit het verband blijkt tusschen corticale hersen-laesie eenerzijsts en diabetes, hemianopsie en vasomotorische stoornis anderzijds. "Deut. Med. Woch.," Oct. 4th, 1888.

HULL (G. S.). Diabetes Mellitus following Abscess of the Gall Bladder. "Med. News," XIII., 1882.

IMLACH (F.). A case of Diabetes Mellitus cured by the removal of the Uterine Appendages. "Brit. Med. Jour." 1885, II., p. 61.

KRATSCHMER. Zur Frage der Glykosurie. "Centralbl. f. d. med. Wissen.," 1886, No. 15, p. 257.

LINDSAY (J. A.). Diabetic Coma, with some remarks on the pathology and treatment of Diabetes. "Dub. Med. Jour.," 1888.

McNISH. Glycosuria complicating a severe burn. "Australasian Med. Gaz.," 1889.

MANBY (A. E.). The connection between Diabetes and Graves' Disease. "Brit. Med. Jour.," 1889, I., p. 1025.

MARSHALL. Quoted by Rollo, *op. cit.*

MAUDSLEY (H.). The Pathology of the Mind. 3rd edition. London, 1879, p. 113.

ORD (W. M.). Remarks on some points of interest in regard to the presence of Sugar in the Urine. "Brit. Med. Jour.," 1889, II., p. 985.

PAVY (F. W.). Introductory Address to the discussion of the Clinical Aspect of Glycosuria. "Lancet," 1885, II., pp. 1033 and 1035.

PEIPER. The Etiology of Diabetes. "Deut. Med. Woch.," 1887, No. 17.

PONIKLO (S. J.). Structural Changes of the Sympathetic Nerve in Diabetes. "Lancet," I., 1878, p. 268.

POULET (V.). Recherches expérimentales sur les phénomènes chimiques de la respiration. "Arch. de Physiol.," 1888, I., p. 174.

PROUT (W.). An inquiry into the nature and treatment of Diabetes, Calculus, and other affections of the Urinary Organs. 2nd edition. London, 1825, p. 60 *et seq.*

REDARD (P.). On ephemeral Glycosuria in surgical practice. Revue de Chir.," 1886, No. 8.

———— Note sur un cas de Glycosurie transitoire au cours abscès chaud. "Arch. Gén. de Méd.," 1888, I., p. 474.

RENDU. Quoted by Debove.

RENZI (E. de). Studii di clinica medica compiuti durante l'anno scolastico 1879-80. "Ann. Univ.," Vol. CCLI.

REUMONT (A.). A case of Tabes Dorsalis complicated with Diabetes Mellitus. "Berlin Klin. Woch.," 1881, No. 13.

RICHARDIERE (H.). Glycosuria and Diabetes occurring in disseminated Sclerosis. "Revue de Méd.," No. VII., 1886.

ROBERTS (SIR W.). A practical Treatise on Urinary and Renal Diseases. 4th edition. London, 1885.

ROBERTSON. Quoted by Campbell.

ROGER (G. H.). Contribution to the study of Glycosuria of Hepatic Origin. "Revue de Méd.," 1886, p. 985.

ROGERS (P. F.). Case of acute Diabetes of unknown duration; death from Coma seventy-two hours after coming under observation. "Boston Med. and Surg. Jour.," Vol. CXI., 1884, p. 298.

ROLLO. *Op. cit.*

SCHEUPLEIN (C.). Verletzung der Wirbelsäule, Diabetes mellitus acutus, vollständige Heilung. "Arch. f. Klin. Chir.," XXIX., p. 865.

SCHMITZ (R.). Meine Erfahrungen bei 600 Diabetikern. Neuenahr. "Proceedings of the Med. Soc.," 1884, p. 279.

——— Kann der Diabetes mellitus ubertragen werden? "Berlin. Klin. Woch., 1890, No. 20.

SCUDAMORE (C.). A Treatise on the nature and cure of Gout. London, 1816, p. 59.

SEEGEN. "Berliner Klin. Woch.," 1887.

SINCLAIR (W. J.). The Glycosuria of Lactation. "Med. Chron.," Vol. III., p. 276.

SMITH (R. SHINGLETON). Remarks on the Morbid Anatomy and Pathology of Diabetes. "Brit. Med. Jour.," 1883, I., p. 657.

SPENCER (W. G.). Diabetes in Surgical Cases. "Westminster Hosp. Rep.," Vol. IV., 1888, p. 89.

STERN. Diabetes in Children. "Lancet," 1890, I., p. 1317.

TAIT (LAWSON). Climacteric Diabetes in Women. "Practitioner," 1886.

TROUSSEAU (A.). Clinique Médicale de l'Hôtel-Dieu de Paris. Vol. II.,p. 771. Clinical Medicine. "New Syd. Soc.," Vol. III., p. 497.

TYSON (J.). A Treatise on Bright's Disease and Diabetes. Philadelphia, 1881.

WEBER (L.). Two fatal cases of Diabetes Mellitus in Children. "American Jour. of Obstetrics," 1884, p. 102.

WHITE (W. HALE). The Pathology of the Central Nervous System in Exophthalmic Goitre. "Brit. Med. Jour., 1889, I., p. 699.

YARROW (H. C.). Temporary Glycosuria, produced apparently by enforced sexual continence. "Philad. Med. and Surg. Rep.," Vol. XLVI., 1882.

ZINN. Ueber Mellituria nach Scharlach. "Jahrb. f. Kinderheilk.," N. F. XIX., 1882.

Chapter III.

MORBID ANATOMY.

(This is a reprint of the Bradshawe Lecture delivered before the Royal College of Physicians in London, August 18th, 1890.)

Mr. President, Fellows of the Royal College of Physicians and Gentlemen :—It is my first duty to express my gratitude to you, and my humble acknowledgment of the honour done to me and I hope I may say to the city to which I belong, in selecting me to deliver the lecture which, by the pious munificence of his widow, annually commemorates the birthday of the late Dr. Bradshawe, a former Member of this College, and a worthy and esteemed physician, who practised during his lifetime in the town of Reading. It is fitting also that I should express on behalf of the College our respectful recognition of this lady's appreciation of the character and aims of our ancient corporation, by founding this lectureship for the promotion of the study of medicine. I feel deeply conscious of my own inability to perform the task so generously entrusted to me in such a manner as to deserve your praise; but by your courtesy I know I may expect a patient hearing, and from your indulgence I trust I may escape your censure if I fail to attain the high level of excellence which has characterised the discourses of my eminent predecessors in this chair.

The subject I have chosen for your attention to-day is undoubtedly wanting in novelty, but I was induced to select it chiefly because it was a matter on which my thoughts were already engaged, and concerning which I possessed abundant materials for a reinvestigation of the questions which have been raised in recent years, with the opportunity of studying them in the light of the most modern methods of histological research. If I have no new discoveries to announce, and if I must content myself rather with the humble position of a critic than that of a revealer of fresh facts or novel theories, I will at least ask you to believe that this has not been from any want of patient industry and careful thought, but because of the great difficulty of making new observations in a field which has engaged the close attention of so many competent observers.

It is my purpose to lay before you an account of the changes observed in the principal organs of the body and in the blood, concluding each description with comments upon points of interest on which my own observations seem to me to have thrown any light.

THE NERVOUS SYSTEM.

Brain.—Though no constant lesion has been observed in the brain, this organ is seldom normal. This statement is at variance with the statistics of cases accumulated from all sources (Windle), but is in harmony with recent observations. Out of twenty-seven cases of which I have records of necropsies the brain was stated to be normal in only five. The most common description given is that it was "œdematous and congested, with thickened membranes," eleven out of the twenty-two abnormal brains presenting more or less of these characteristics. It is less often described as anæmic. Atrophy of the convolutions has been stated to be common (Mackenzie); but this does not accord with my observations. These changes are, of course, not peculiar to diabetes, and cannot be considered to have any special relation to its pathology. In quite a minority of cases localised lesions are found, the value of which differs greatly, and in many instances is not easily appraised. Among the most important are tumours in the fourth ventricle and medulla, of which a relatively small number only have been recorded. Not a single instance has come under my own observation, and I have only found records of ten cases, of which three were published by Frerichs. Of the direct dependence of the diabetes upon the growth in many of these cases there can be no reasonable doubt. Instances have also been recorded of softening (Luys); of the presence of corpora amylacea and colloid masses (Abraham); of sclerosis; of alterations in colour and congestion (Tardieu); and of enlargement of the peri-vascular spaces (Dickinson). Of these the softening alone is of undoubted value. The congestion and dark colour of the medulla are interesting, but, taken by themselves, prove little; while the enlargement of the peri-vascular spaces is too common in non-diabetic brain disease, and too exceptional in diabetes, to count for much. The observations of Dr. Abraham are a valuable contribution to the histology of the diabetic brain, and possibly belong to the same class as other lesions, to which I shall refer immediately, which indicate failure of nutrition of the brain tissues, and are rather an effect than a cause of the disease.

There is undoubtedly a tendency in the diabetic brain to the formation of cysts in the white matter. I have found such cysts in the frontal lobes, in the medulla, and in the pons; while in another case a small focus of softening in each crus cerebri looked like an early stage in this process of cyst formation. These cysts are quite free from hæmatoidin staining. Some of them are so small as to be hardly worthy of the name of cysts, but others are as large as horsebeans. It seems probable that this condition is due to a failure of nutrition. The choroid plexuses also occasionally present abnormalities; thus I have found cysts on the choroid plexus of the left lateral ventricle, and the same structure in the fourth ventricle was in one case hypertrophied, while in another case the plexuses of the lateral ventricles were of a dark-purple colour, as if from congestion. The lateral ventricles and the *iter a tertio ad quartum ventriculum* have been found dilated without any mechanical obstruction to explain it. Extensive hæmorrhage into the brain substance is rare in diabetes. In Windle's tables there are only three examples in one hundred and eighty-four necropsies, or in 1·6 per cent. There has been no instance at the General Hospital for twenty years out of a total number of one hundred and twenty cases with twenty-nine deaths. Dickinson has, however, described minute hæmorrhages as common. Finally, glycogen has been found in large quantities in the medulla oblongata and in the vessels of the cerebral cortex (Futterer). Zaleski found iron present, but we are still in doubt as to how far this is an abnormality.

On careful microscopical examination I have not been able to detect, even with the most modern technical methods, any special histological changes either in the cortex, the basal ganglia, or the medulla. The latter has been the subject of special investigation, but the only positive fact observed has been that the capillaries of the vagus nucleus in one case seemed to be abnormally numerous and full of blood. The specimen has been placed under one of the microscopes. I have looked for minute hæmorrhages without finding them, and the absence of blood pigment in the cysts already alluded to is opposed to the view that these originate in that fashion. While not disputing its occasional occurrence, hæmorrhage is certainly not a constant or common form of lesion in the diabètic brain.

Spinal Cord.—In a certain number of cases diabetes has followed the extension of diseased processes from the spinal cord into the medulla, as in locomotor ataxy and insular sclerosis, where the disease is unquestionably the result of the lesion of the medulla. Again, diabetes has not infrequently followed injuries to the spinal column, though in one case where this occurred the cord was apparently uninjured. In connexion with these facts, it will be remembered that Schiff has produced artificial glycosuria by dividing the cord opposite the second dorsal vertebra; but in cases where the diabetes is neither a complication of a recognised disease of the cord nor a consequence of injury to the spine, this structure has been usually described as normal. Unfortunately, the number of cases in which it has been carefully examined is not large, but in these enough has been found to shew that secondary nutritive changes, similar to those in the brain, are apt to occur in the cord; these are dilatation of the central canal, enlargement of the peri-vascular sheaths, and localised softening. Tumours of the cord in connexion with diabetes have been very rarely recorded; the only case known to me is one of myxoma of the dura mater (Shingleton Smith). Glycogen has been found in large quantities in the spinal cord. Microscopical examination of the spinal cord, stained after Weigert's method, has been carefully carried out without any special change being noted.

Cerebro-spinal Nerves.—Tumours situated on or compressing the vagus nerve have been found associated with diabetes. Three such cases have been recorded (Harley, Henrot, Frerichs), and in each instance the right nerve was the seat of the lesion. In Frerichs' case the tumour encroached upon the floor of the fourth ventricle, but in the other two cases no such complication existed, the tumour being situated in the thorax. These observations are very interesting in connexion with the recent experiments of Arthaud and Butte, who found that artificially induced neuritis of the central end of the divided vagus caused glycosuria, while a similar lesion of the peripheral end caused hunger, wasting, polyuria, and thirst. Lubimoff has found in one case of diabetes atrophy and pigmentation of the inferior ganglion of the vagus. No pathological observations on the spinal nerves in diabetes are known to me, but it is noteworthy that Arthaud and Butte have produced glycosuria by setting up neuritis in the roots of the first dorsal pair of spinal nerves. Schiff, moreover, has shewn that stimulation of the central end of any sensori-motor nerve, such as the sciatic, may be followed by glycosuria. Clinical observation has revealed the existence of a secondary diabetic neuritis, which may be multiple (Leyden), or attack particular nerves (Althaus). In these circumstances it gives rise to the phenomena of paralysis, with wasting and loss of faradaic response of the muscles supplied by the affected nerves.

Sympathetic Nerves.—Changes in this system of nerves in diabetes seem very early to have attracted the attention of pathologists. Thus Duncan, as long ago as 1818, found the sympathetic in the abdomen three times as thick as normal; and Percy, in 1842, described the semilunar ganglia, the splanchnic nerves and vagus, as thickened and of cartilaginous hardness. Klebs and Ph. Munk, in 1870, found changes in the cœliac plexus with destruction of a number of ganglion cells. Lubimoff has also found sclerosis of the sympathetic ganglia and atrophy of their nerve cells. In three of my cases the semilunar ganglia have been found enlarged, and in one case atrophied, with increase of connective tissue and atrophy of nerve cells. Hale White has recently described similar lesions in four cases of diabetes. On the other hand Shingleton Smith has made numerous observations on the state of the sympathetic ganglia without finding any uniform or definite change. In three of my cases the semilunar ganglia were normal, and I have shewn in a paper published in the *British Medical Journal* in 1883 that similar microscopical changes to those above described are met with apart from diabetes. Yet

I have never seen the semilunar ganglia enlarged except in diabetes; in some of my cases the right semilunar ganglion was quite twice the normal size. The importance of these facts depends upon the results of experiments which have shewn that destruction of various sympathetic ganglia—for example, the superior and inferior cervical (Pavy), the first thoracic (Eckhard), and the abdominal (Klebs),—and division or ligature of the splanchnic nerves (Hensen, Arthaud and Butte), are followed by glycosuria. Extirpation of the cœliac plexus is followed by wasting and death, with temporary glycosuria and acetonuria (Lustig, Peiper).

THE CIRCULATORY SYSTEM.

Heart.—In about 40 per cent, of my cases the heart has been described as free from noticeable change; while in about the same proportion it has been described as

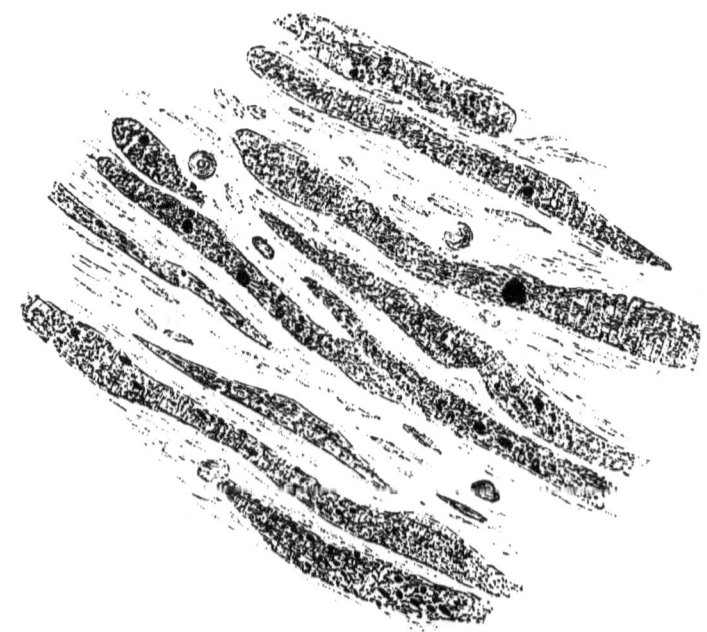

Fig. 3. Heart muscle from a case of very advanced fatty and fibroid degeneration. Fat droplets stained black with osmic acid. (Hartnack oc. 2, obj. 7. Tube drawn out.)

pale and soft; more rarely, dilated or hypertrophied, or distinctly fatty. Pericarditis occurs occasionally, and, in one case of death from a carbuncle in a diabetic subject, the pericardium was full of fluid, fatty blood. Valvular disease is quite exceptional, though Lecorché has described endocarditis as a complication of diabetes, and Maguire has recorded a case. It is said to occur in the later stages, and to affect usually the auricular surface of

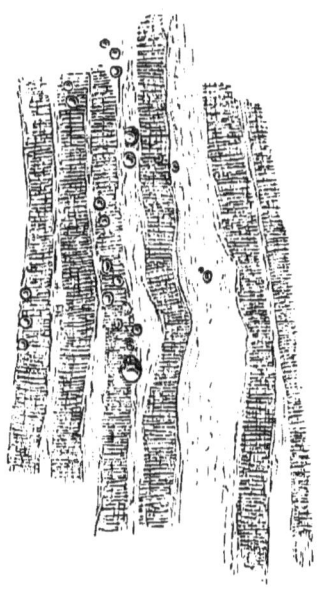

Fig. 4. Heart muscle apparently healthy, except for a few granules of glycogen lying within the sheaths of the muscular bundles. (Hartnack, oc. 2, obj. 7. Tube drawn out.)

the mitral valve. The rarity of endocarditis may be estimated from the occurrence of only one case of valvular disease out of the ninety-four cases collected in Windle's tables; and in my twenty-nine cases there was valvular thickening in one only. Mayer has stated that out of three hundred and eighty diabetics seen at Carlsbad, eighty-two, or 24 per cent., shewed signs of cardiac enlargement, and Schmitz, of Neuenahr, has made a still stronger statement as to the prevalence of fatty heart. Frerichs has drawn attention to glycogenic degeneration of the cardiac muscle, and suggests that this is the real cause of the cardiac debility of diabetes. Hypertrophy of the heart was present in 13 per cent. of cases examined in the Berlin Pathological Institute, and this accords very nearly with my experience. Fatty heart is probably much more common, if we are to judge by the frequency with which the muscular substance is described as pale and soft in reports of diabetic necropsies; yet I should estimate it as occurring in less than 40 per cent. of all cases. The preceding illustration (*Fig.* 3) shews the appearance of the muscle in a very marked case. Glycogenic deposits in the wall of the heart have only been found by me in such small quantity that I cannot ascribe to them any serious pathological significance. The drawing (*Fig.* 4) shews a few granules of glycogen deposited between the muscular bundles, but the muscular fibre appears healthy.

Blood.—The blood of diabetics generally looks normal to the naked eye. It is sometimes described as dark, and is sometimes very obviously loaded with fat, a white, cream-like layer rising to the surface when the blood is allowed to stand. Under the microscope the fat is seen to be at first in a state of molecular subdivision, but these molecules run together to form droplets after death, and this may produce the appearance of capillary embolisms. Occasionally the red blood-corpuscles are found broken down into a granular material (Foster, von Jaksch). In one of my cases the leucocytes were peculiarly large. Quantitative changes in the hæmocytes are common, these being generally reduced in number, with a corresponding reduction in the hæmoglobin. Normal blood contains sugar in varying amounts. The following table gives the estimate of four observers:—

TABLE.

Name.	Parts per cent.
Pavy	0·078 to 0·081 (dog).
Otto	0·10 ,, 0·14 ,,
Seegen	0·15 ,, 0·19 (man).
Frerichs	0·12 ,, 0·3 ,,

According to Seegen, the amount present in mild cases of diabetes does not exceed the normal, but in severe cases it may rise as high as 0·4 parts per cent. The alkalinity of the blood serum is reduced, owing to the presence of certain organic acids of doubtful identity, of which diacetic acid and β-oxybutyric acid seem the most probable. The existence of acetone is disputed, as former reports of its presence are believed to be due to the breaking up of diacetic acid to form acetone and carbonic acid gas. In certain experiments conducted by me some years ago in the laboratory of Professor Tilden, at Mason College, I was unable to detect any acetone in the blood of a diabetic patient who had died of coma.

THE RESPIRATORY SYSTEM.

Lungs.—Pathological alterations in the lungs are the rule in diabetics, and perhaps no organ shews more constant changes. In my cases 17 per cent. only were free from disease. The most common condition was congestion, or congestion and œdema. The next most frequent alteration was phthisis, which was present in 27 per cent. Small foci of softening were observed in one case, abscess in one, hæmorrhagic infarcts in one, and gangrene also in one. Dreschfeld has described the following types of lung affection in diabetes: (1) Acute croupous pneumonia, very acute and fatal, but rare; (2) acute broncho-pneumonia, which may terminate by gangrene; (3) chronic caseating tubercular broncho-pneumonia, the common form of diabetic lung complication; (4) chronic non-tubercular broncho-pneumonia; (5) gangrene of the lung. Fink recognises two forms of diabetic phthisis: (1) Tubercular and (2) fibroid. In the latter there are no tubercle bacilli in the sputa of lung, no caseous deposits, and the lung undergoes chronic induration. Pleurisy and empyema occur rarely. Fat embolisms have been described as playing an important part in the pathology of diabetic coma (Sanders and Hamilton), but it is doubtful if they are not post-mortem formations, due to the running together of the fat which was previously held suspended in a molecular state, and in any case they are not present in such numbers as to give rise to any symptoms (Saundby and Barling). The vessels of the lungs have been described as undergoing hyaline and fibroid thickening; but this is not a primary change, or one in any way peculiar to diabetes.

THE DIGESTIVE SYSTEM.

Liver.—Great interest attaches to this organ, because of its physiological relations to sugar formation. An opinion was at one time held, and has been expressed by recognised authorities, that the liver in diabetes is usually healthy, and even a more than usually healthy appearance has been described as characteristic of it. In my experience this is a very erroneous view. The liver is generally enlarged, weighing from sixty to eighty ounces. In a smaller number it is small, pale, and soft. Fatty degeneration is very common; congestion is often observed; the consistence of the organ is sometimes abnormally firm. A certain degree of interstitial hepatitis is frequently present, and occasionally this goes so far as to produce distinct cirrhosis. This is attributed by Letulle to the effect of the abnormal destruction of hæmocytes. Hanot and Schachman have also described this form of cirrhosis. The liver is sometimes smooth, at others granular and scarred. The lesion begins, according to these observers, around the radicals of the hepatic vein, but Brault and Gaillard point out, in my opinion rightly, that the new growth begins in both the hepatic and portal areas. This form of cirrhosis is associated with bronzing of the skin. Some degree of interstitial hepatitis may, however, be seen without any evidence of pigmentary deposit in the liver or the integument. (*Fig.* 5.) This drawing shews very well marked com-

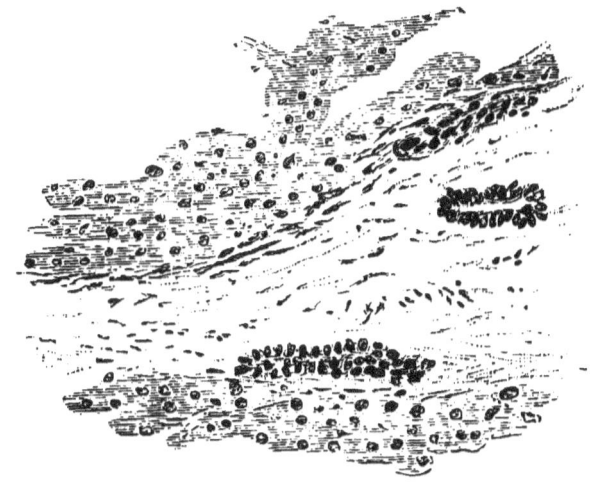

Fig. 5.— Section of liver shewing commencing interstitial hepatitis in a portal canal. Two newly formed biliary canaliculi are seen near the edges of the acini. (Hartnack, oc. 3, obj. 7 Tube drawn out.)

mencing cirrhosis in a diabetic liver. Abscess of the liver is sometimes met with associated with diabetes. I remember one such case when I was a pathologist, and I believe the case presented all the phenomena of severe diabetes. The liver contained one large abscess and numerous smaller ones. A similar case was reported in the U.S. Navy Reports for 1878. It is permissible to believe that the abscess in these instances was really the cause of the diabetes. Dickinson has described thrombosis of the branches of the portal vein, and angeiomata formed of dilated capillaries near the radicals of the hepatic vein. In spite of its fatty appearance Weyl and

5

Apt have not found the diabetic liver to contain an excessive amount of fat:—

 Normal liver · · · 3·70 pts. per mille.
 Diabetic liver · · · 3·75 ,, ,,

Moreover, absence of fat from the liver cells has been observed by Beale and Frerichs. Quincke described excess of iron in the liver, deposited in the form of granules. Zaleski has not observed it in granular form; he estimates the diabetic liver to contain 0·685 parts of iron per mille, but he points out that we have no data to enable this to be compared with the normal quantity.

Spleen.—This organ is very frequently described as normal, but the most common naked-eye change is that it is small, pale and soft. It is more rarely enlarged and congested, and sometimes contains tubercle. It is said to contain excess of iron (Quincke), but, as already explained, data are wanting as to the amount present in health. Glycogen has also been found in it. Hyaline degeneration of the small arteries has been described, but I have not been able to perceive anything abnormal about them.

Pancreas.—Since Lancereaux drew attention to the frequency with which the pancreas is atrophied in diabetes, and went so far as to associate the clinical type of *diabète maigre* with this lesion, the changes in this organ have been interesting; but they are especially so since the experiments of Minkowski, Lépine, and others have shewn that extirpation of the pancreas in animals is followed by glycosuria. This organ is by no means so carefully examined by pathologists as it deserves to be, and there is wanting that familiarity with its ordinary naked-eye and microscopic appearances which gives value to their descriptions. In many cases no note of its condition has been made. Out of fifteen cases it was atrophied in seven and abnormally firm or fibroid in four, while it was normal in only four. In two of the cases in which it was hard it was also enlarged. So far as my own observations go, I am disposed to agree with Lancereaux, as I have found the pancreas shrunken in all my cases of typical wasting diabetes. That the pancreas stands in some relation to diabetes is shewn by the experience of Bull, of New York, whose patient died of diabetes after he had extirpated the pancreas. Duffey has published a case of diabetes associated with cancer of the pancreas; on the other hand, in the cases of pancreatic disease collected by Dr. Handfield Jones, and more recently by another distinguished Fellow of this College, Dr. Norman Moore, there is no mention of diabetes. [Dr. Baumel, in a paper published in the "Montpellier Médical" for 1881-82 was the first to contend that disease of the pancreas was the regular cause of diabetes, and in the same communication he recorded a case of diabetes without emaciation *(diabète gras)* in which this association existed. To him, therefore, belongs by right any honour that is due to the writer who first distinctly recognised the full significance of the pancreatic lesion in diabetes. In a recent article (August 2nd, 1890) the editor of the "Philadelphia Medical News" refers to a case of cystic disease of the pancreas from a case of diabetes exhibited by Longstreth in 1877, and to one of multiple pancreatic abscesses with glycosuria reported by Frison in 1875; on the other hand, he states that Langenhans has recently published a case of necrosis of the pancreas in which there was no sugar in the urine.] Cyr thinks the atrophy is secondary, and only a result of the general failure of nutrition. Ligature of the duct, and even extirpation of the gland, have not always been followed by diabetes; but the force of this objection is removed by the observation that a small part of the pancreas, if left behind, is capable of supplying the necessary ferment (?). Various theories have been put forward to account for the action of the pancreas.

It has been suggested, (1) that the pancreas supplies the liver with the fatty acids necessary to the formation of bile acids from glycogen, hence, in affections of the pancreas there is excess of glycogen with resulting diabetes (Popper); (2) excessive production of glucose from diastatic ferments formed in the stomach and duodenum (Bouchardat); (3) the formation of paraglucose by the

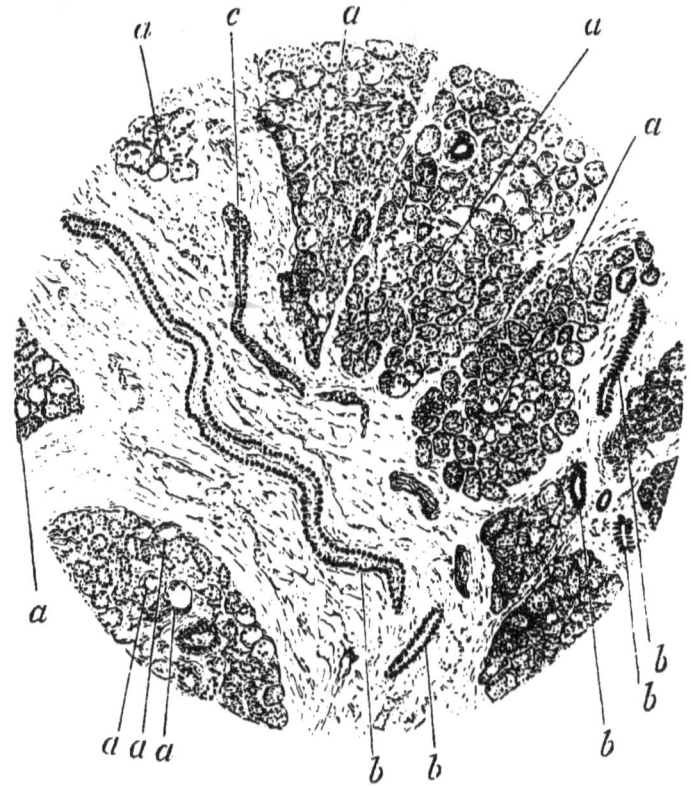

Fig. 3.—Section of atrophic cirrhotic pancreas, shewing great increase of connective tissue with *a*, hyaline degeneration of glandular epithelium; *b*, newly formed ducts and *c*, ducts undergoing atrophy. (Hartnack, oc. 2, obj. 4. Tube drawn out.)

action of modified pancreatic juice (Cantani); (4 the absorption of the pancreatic ferment, which, passing direct to the liver, changes the glycogen to glucose (Baumel); and (5) the suppression of a normal sugar-destroying ferment which should pass by the pancreatic lymphatics to mingle with the chyle (Lépine). Brault and Gaillard have described a pigmentary cirrhosis of the pancreas analogous to the pigmentary cirrhosis of the liver observed in diabetes. Pancreatic calculi may cause cystic formation or sclerosis, or complete adipose transformation.

The drawing (*Fig*. 6) is a low-power view of a portion of pancreas, shewing great increase of connective tissue with a new formation (?) of ducts and a hyaline degeneration of the pancreatic glandular epithelium. The portions of gland affected by this last lesion seem to become converted into swollen convoluted masses of translucent

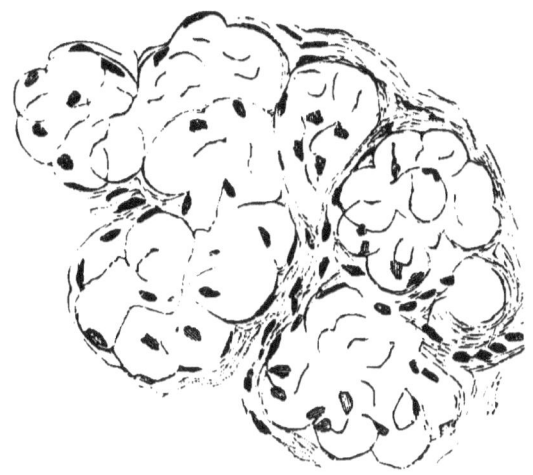

Fig. 7.—Highly magnified view of a group of pancreatic acini which have undergone hyaline degeneration. (Hartnack, oc. 3, obj. 7. Tube drawn out.)

material containing a few pale nuclei, and bounded by a layer of flattened endothelial cells, whose nuclei seen edgeways look like spindle cells (*Fig*. 7). It is evident that there are two processes going on in the gland: (1) a degeneration of the gland elements, closely resembling that to be described in the epithelium of the kidneys, which is apparently peculiar to diabetes; and (2) an increase of fibrous tissue with destruction of the gland analogous to cirrhosis of the liver. This process is undoubtedly inflammation and not simple atrophy. In the earlier stages of the process the gland is swollen from abundant infiltration of its tissue with round cells.

A section of such a swollen pancreas, obtained from a case of rather acute diabetes, is shewn under one of the microscopes (*Fig.* 8). It has been suggested that the

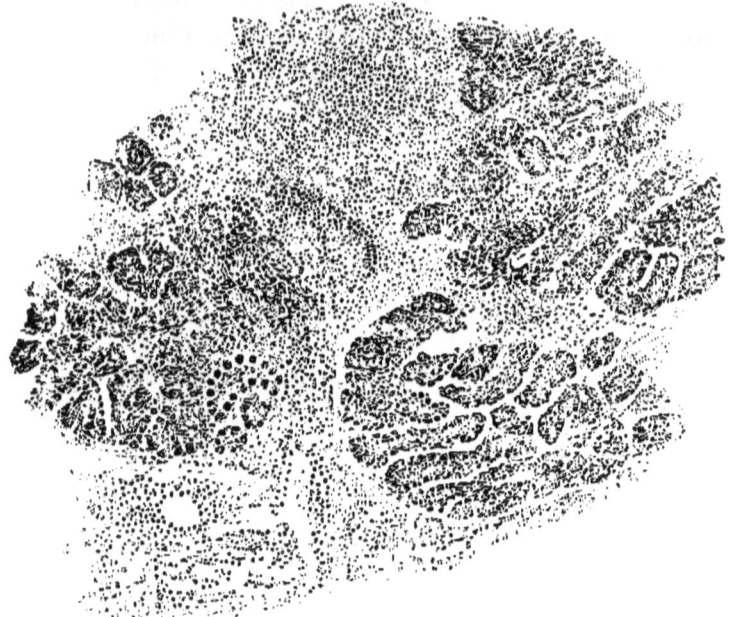

Fig. 8.—Section of enlarged pancreas shewing commencing cirrhosis with infiltration of the connective tissue spaces by round cells. (Hartnack, oc. 2, obj. 4. Tube drawn out.)

atrophy of the pancreas may be secondary to disease of the cœliac plexus; but Lustig found that extirpation of this plexus was not followed by atrophy of the pancreas; and this has been confirmed by Peiper.

Stomach.—It is unfortunate that the stomach is often omitted in post-mortem reports. I have notes of its condition in only eleven cases, but it is significant that in only one instance was it found to present no morbid alteration. In three cases it was dilated, in three there were hæmorrhages generally near the pyloric end, in two the mucous lining was congested, in one softened, and in one in a state of chronic catarrh. Windle speaks of "thickening" of the mucous membrane as not uncommon, and also of distension of the organ by gas.

Intestines.—The large bowel is generally found filled with hardened fæcal masses, and congestion or catarrhal conditions of the mucous membrane are not uncommon. Hæmorrhages similar to those in the stomach also occur in the duodenum. In two instances large masses of tænia medio-canellata were present. The mesenteric glands were twice found to be enlarged. Frerichs has described a dysenteric condition of the large intestine, and Ebstein desquamation of the epithelium lining the bowel.

GENITO-URINARY SYSTEM.

Kidneys.—In not one of my cases were the kidneys described as normal, though in many the changes were not of great importance. The most common condition appears to be a slight degree of fatty degeneration. This comes out in Windle's tables as well as my own. Such kidneys are generally enlarged, and their capsules are often adherent. Less commonly the kidneys are enlarged and congested. Sometimes there is thinning of the cortex, or distinct contraction of the organ. Tubercle and lardaceous disease occur occasionally, and the kidney may become gangrenous (Turner). The above-mentioned fatty change has been described specially by Fichtner, who regards it as characteristic of diabetes. But the most interesting histological fact in the diabetic kidney is the hyaline degeneration of the tubular epithelium, which was first described by Armanni. His observations were confirmed by Ebstein, who assigned the seat of the lesion to the descending limb of Henle's tubes; but this was disputed by Ferraro, who, however, found the change nowhere outside the medullary portion of the kidney. This last observer attributed the changes to a primary lesion of the vascular walls. These changes in the epithelium and the co-existing hyaline changes in the vessels were well described by a distinguished Fellow of this College, Dr. Stephen Mackenzie, in opening a discussion at the Pathological Society of London in 1883. Ehrlich also described, " as characteristic of and peculiar to diabetes," a hyaline degeneration of the epithelium of that part of the kidney between the cortex and the

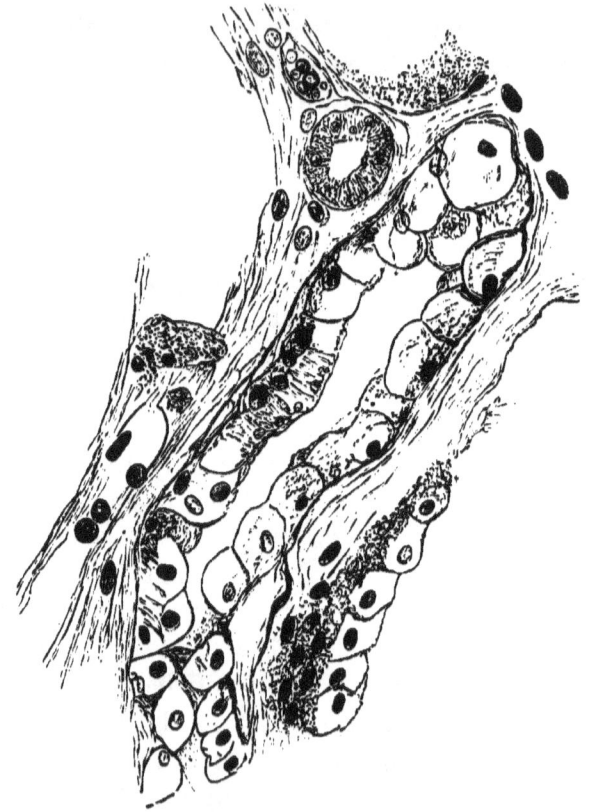

Fig. 9. Portion of Henle's looped tube, shewing hyaline degeneration of the epithelium Armanni's lesion. (Hartnack, oc. 2, obj. 7. Tube drawn out.)

pyramid—that is, the boundary zone; and Strauss proved that this lesion is identical with that described by Armanni. Strauss found that it was only present in one out of three cases, but he agreed that it is absolutely characteristic of diabetes, and he attributed it to the osmosis of sugar into the cells. Ehrlich considered that the change was due to glycogenic infiltration of the cells, but Strauss, in a later paper, says that no glycogen may be visible, even when all Ehrlich's precautions are taken.

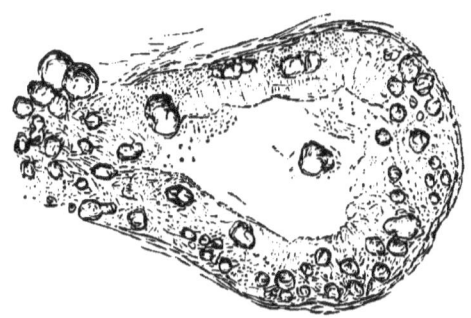

Fig. 10. Renal tubule shewing masses of glycogen within the epithelial cells or free. (Hartnack, oc. 3, obj. 7. Tube drawn out.)

With reference to its causation, Albertoni and Pisenti have investigated the changes produced by administering small doses of acetone to rabbits. They found the following lesions : (1) granular degeneration of the epithelium of the convoluted tubes, the nuclei still staining well ; (2) complete destruction of the epithelium with disappearance of the nuclei, and slight dilatation of Bowman's capsules; the glomeruli and connective tissue were not affected. They suggest the possibility that the necrosis of the epithelium observed by Ebstein may be due to the action of acetone on the kidneys. Stokvis, too, found that artificial diabetes, produced by injecting grape sugar into the blood, caused albuminuria ; and he believes that this sets up a chronic nephritis. Both these are possible explanations of the fatty change already alluded to, which is entirely like that of slight chronic parenchymatous nephritis due to infective disease; but they do not throw any light upon this peculiar and characteristic hyaline change. My observations entirely confirm the existence of this hyaline degeneration and its restriction to the epithelium of Henle's tubes. It is well shewn in the illustration (*Fig.* 9), and is quite different from the glycogenic infiltration which may also be present, as shewn in the drawing (*Fig.* 10), though this fails to bring out the characteristic mahogany staining of the glycogen granules. The kidneys, like many other organs in diabetes, may present marked fatty degeneration (*Fig.* 11),

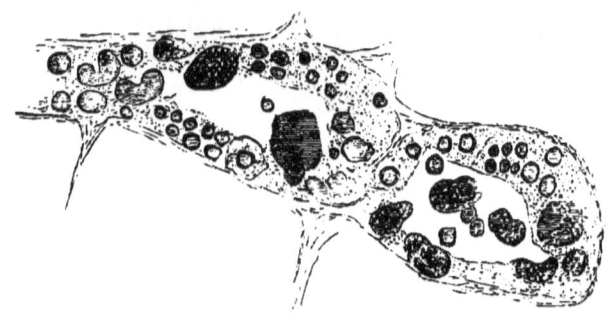

Fig. 11. Portion of renal tubule shewing fat droplets, stained black with osmic acid, lying within the epithelial cells or free in the lumen. (Hartnack, oc. 3, obj. 7. Tube drawn out.)

droplets staining black with osmic acid being abundantly visible within the epithelial cells and free in the lumina of the tubules, the straight tubules being chiefly affected. Not uncommonly the kidneys present all the characteristic features of chronic diffuse nephritis, as is illustrated by one of the microscopic specimens.

Bladder.—This organ is usually normal, but it is often dilated and hypertrophied. The mucous lining sometimes may be the seat of hæmorrhages.

Ovaries.—Israel has described a case in which there was gangrene of both ovaries associated with recent general peritonitis. These organs have also been described as cystic and atrophied, but these are common conditions, and not in any way specially associated with diabetes.

Concluding remarks—It is very plain that diabetes, far from having no morbid anatomy, has one of a very complicated kind, and one that cannot be without bearing upon its pathology. The most important lesion is, in my opinion, the wasting of the pancreas; but it is not until this organ has been carefully examined in a larger series of cases than I have been able to collect that we shall be in a position to speak definitely as to its pathological position and value. Of great importance, too, are the changes in the abdominal sympathetic ganglia, though they are too inconstant to form the basis of a satisfactory theory. The liver changes are probably altogether secondary to a functional hyperæmia induced under nervous influence. The other changes in kidneys, lungs, heart, and brain are merely the results of defective nutrition and of long-standing hyperglycæmia. But the morbid anatomy of diabetes is not yet completed, as is shewn by our imperfect knowledge of the state of the pancreas. Unfortunately diabetes is a rare disease even in our great hospitals, hence it is a work of many years to accumulate a series of carefully observed cases, so that with our custom of holding pathological appointments only for a few years, few of us get the opportunity of doing so. I have to express my thanks to the late Pathologist to the General Hospital, Dr. G. F. Crooke, and to his successor, Dr. Ratcliffe, for the assistance they have given me in this work in many ways, and especially in cutting and mounting microscopic sections.

BIBLIOGRAPHY.

ABRAHAM (P. S.). On some Microscopical Lesions from two cases of Diabetes Mellitus. "Dub. Med. Jour.," Vol. LXXVII., p. 395.

ALBERTONI and PISENTI. Azione dell' Acetone e dell' Acido Acetacetico sui Reni. "Arch. per le Sc. Med.," 1887, Vol. XI., pp. 129 to 152.

ALTHAUS. Neuritis of the Circumflex Nerve in Diabetes. "The Lancet," Vol. I., 1890, p. 455.

ARMANNI and CANTANI. Le Diabète Sucré et son Traitement Diététique. Paris, 1876.

BRAULT and GAILLARD. Sur un cas de Cirrhose Hypertrophique dans le Diabète Sucré. "Arch. Gén. de Méd.," Vol. I., 1888.

CYR (J.). Etude Critique sur quelques Travaux Récents concernant l'Anatomie Pathologique du Diabète. Paris, 1880.

DICKINSON (W. H.). On Diabetes. London, 1877.

DRESCHFELD (J.). On the Pathology of the Lung Complications in Diabetes. "Med. Chronicle," Vol. I., p. 5.

DUFFEY (G. F.). On the Connection of Acute Diabetes with Disease of the Pancreas. "Trans. Acad. Med. in Ireland," Vol. II.

DUNCAN. Clinical Reports, 1888, case 28, p. 137.

EBSTEIN (W.). Ueber Drusen-Epithelnekrosen beim Diabetes mellitus mit besonderer Berücksichtigung des diabetischen Coma. "Deutsches Arch. f. klin. Med.," 1881, Bd. XXVIII., p. 143.

————. Weiteres über Diabetes mellitus insbesondere über die Complication derselben mit Typhus abdominalis. *Ibid.*, 1882, Bd. XXXI., p. 1.

EHRLICH. Ueber das Vorkommen von Glycogen in diabetischen und normalen Organismus. "Zeit. für klin. Med.," 1883, Bd. VI., p. 33.

FERRARO (P.). Nuove Ricerche sulle Alterazioni degli Organi nel Diabete Mellito. "Il Morgagni," 1883, pp. 71 and 233.

FICHTNER (R.). Zur pathologischen Anatomie der Nieren beim Diabetes mellitus. "Virchow's Archiv.," Bd. CXIV.

FINK (H.). Zur Lehre von den diabetischen Lungenerkrankungen. "Münch. Med. Woch.," 1887, No. 37.

FOSTER (B. W.). Diabetic Coma—Acetonæmia. "Brit. Med. Jour.," 1878. Vol. I., p. 78.

FRERICHS (T.). Ueber den Diabetes. Berlin, 1884.

————. Ueber den plötslichen Tod und über das Coma beim Diabetes (Diabetische Intoxication). "Zeit. für klin. Med.," Bd. VI., Heft. 1.

FUTTERER (G.). Glykogen in der Capillaren der Grosshirnrinde beim Diabetes mellitus. "Cent. f. d. Med. Wiss." 1888, No. 88.

GUILLIOT (O.), Glycosurie et Inosiurie; Dégénérescence Graisseuse du Pancreas. "Gaz. de Paris," 1881.

HANOT and SCHACHMANN. Sur la Cirrhose Pigmentaire dans le Diabète Sucré. "Arch. de Physiol.," 1886, Vol. I., p. 50.

HARLEY (quoted by ROBERTS). Urinary and Renal Diseases.

HENROT. "Bull. de la Soc. Méd. de Rheims," 1874, No. 13.

HUGOUNENQ. Contribution à l'Etude de la Dyscrasie Diabétique. "Rev. de Méd.," 1887, p. 301.

ISRAEL (O.). Zwei Fälle von Nekrose innerer Organe beim Diabetes mellitus. "Virchow's Archiv.," Bd. LXXXIII., 1881.

JAKSCH (R. V.). Ein Fall von Sogen Acetonämie. "Prag. Med. Woch.," 1880.

JONES (C. HANDFIELD). Observations respecting Degeneration of the Pancreas. "Med. Chir. Trans.," Vol. XXXVIII., 1855.

KLEBS and PH. MUNK. "Handbuch der Path. Anat.," Bd. III., 1870, p. 547.

—————————. "Tagebl. der Innsbrücker Naturforscher Vers.," 1869, p. 113.

LANCEREAUX (E.). Notes et Réflexions sur un cas de Diabète Sucré avec altération du Pancréas. "Bull. de L'Acad. de Méd.," 2e. Sér., VI., 46, 1887.

—————————. Le Diabète Maigre et le Diabète Gras. "L'Union Méd.," 1880, XIII., 16.

LECORCHÉ. De l'Endocardite Diabètique. "Arch. Gén. de Méd.," 1882.

—————————. Traité du Diabète. Paris, 1877.

LETULLE. Deux cas de Cirrhose Pigmentaire dans le Diabète Sucré. "Soc. Méd. des Hôpitaux," 1885.

LEYDEN. Die Entzündung der peripheren Nerven. Berlin, 1888.

LUSTIG (A.). Sugli Effetti dell' Estirpazione del Plesso Celiaco. "Arch. per le Scienze Med.," 1889, No. 6.

LUYS. "L'Encéphale," May, 1882.

MACKENZIE (S.). Discussion upon the Morbid Anatomy of Diabetes." "Trans. Path. Soc.," 1883.

MAGUIRE (R.). Roberts' Renal and Urinary Diseases. 1885, p. 271, foot note.

MAYER (J.). Ueber den Zusammenhang des Diabetes mellitus mit Erkrankungen des Herzens. "Zeit. f. Klin. Med.," Bd. XIV., Heft 3.

MOORE (NORMAN). Cancer of the Pancreas. "St. Barth. Hosp. Rep.," Vol. XVII., 1881, p. 205.

NORTA. Observation du Diabète Maigre. "L'Union Méd.," 1881.

OTTO (J. G.). Ueber den Gehalt des Blutes an Zucker und reducirender Substanz unter verschiedenen Umständen. "Pflüger's Archiv.," 1885, Bd. XXXV., p. 467.

PAVY (F. W.). Croonian Lectures on Points connected with Diabetes. "The Lancet," 1878, Vol. I., p. 557.

PEIPER. Experimental Studies on the Results of Extirpation of the Cœliac Plexus. "München. Med. Wochen.," April 29th, 1890.

PERCY (J.). Case of Diabetes Mellitus. " Med. Gaz.," 1842 to 1843, Vol. I., p. 49.

PILLIET (A.). Scléroses du Pancréas et Diabète. " Le Prog. Méd.," 1889, No. 21, p. 391.

PONIKLO (S. J.). *Op. cit.*

QUINCKE (H.). Ueber Siderosis, Eisenablagerung in einzelnen Organen des Thierkörpers. " Fest. zum And. v. Al. v. Haller," Bern, 1877.

SANDERS and HAMILTON. Lipæmia and Fat Embolism in the Fatal Dyspnœa and Coma of Diabetes. " Edin. Med. Jour.," Vol. XXV., Part 1, 1879, p. 47.

SAUNDBY and BARLING. Fat Embolism. " Jour. of Anat. and Phys.," Vol. XVI., 1882, p. 515.

SCHIFF (J. M.). Untersuchungen über die Zuckerbildung in der Leber. Würzburg, 1859. " Jour. de l'Anat. et de la Phys.," Vol. III., p. 369.

SCHMITZ (R.). On a Frequent and Noteworthy Complication of Diabetes mellitus. " Lond. Med. Record," Vol. IV., 1876, p. 200.

SEEGEN (J.). Ueber die Zuckerbildung in der Leber. " Arch. f. Anat. und Physiol.," 1880.

—————. Ueber dem Zuckergehalt des Blutes vom Diabetikern. " Wien. med. Wochen.," 1886, No. 47.

SMITH (R. SHINGLETON). Remarks on the Morbid Anatomy and Pathology of Diabetes. " Brit. Med. Jour.," 1883, Vol. I., p. 657.

STRAUSS (J.). Contribution à l'Etude des Lésions Histologiques du Rein dans le Diabète Sucré. " Arch. de Phys.," 1885, Vol. II., p. 322.

—————. Nouveaux Faits pour servir à l'Histoire des Lésions Histologiques du Rein dans le Diabète Sucré. *Ibid.*, 1887, p. 76.

TARDIEU (quoted by ROBERTS). *Op. cit.*

TURNER. Necrosis of the Pyramids of the Kidney. " Brit. Med. Jour.," 1888, Vol. I., p. 1059.

WEYL and APT. Ueber den Fettgehalt kranker Organe. " Virchow's Archiv,," Bd. XCV., p. 351.

WHITE (W. HALE). On the Sympathetic System in Diabetes. " Path. Soc. Trans.," Vol. XXXVI., p. 67.

WINDLE (B. C. A.). The Morbid Anatomy of Diabetes Mellitus. " Dublin Jour. of Med. Science," Vol. LXXVI., 1883, p. 112.

ZALESKI (S.). Zur Pathologie der Zuckerharnruhr (Diabetes mellitus) und zur Eisenfrage. " Virchow's Archiv.," Bd. CIV., p. 91.

Chapter IV.
CLINICAL HISTORY.
SYMPTOMS.

DIABETES Mellitus presents itself for clinical observation under two types, (1) acute, (2) chronic, in which the essential symptoms, *wasting, thirst, polyuria* and *glycosuria* are present in varying degrees of intensity. In the former these are all great, in the latter the first three especially are moderate, and the total excretion, though perhaps not the actual percentage of sugar, is also not excessive.

In the acute type the patient is young, he looks ill, his face is pale or cyanosed, his expression is anxious, the hair is rough, the skin dry, the lips parched, the tongue red and sticky, or covered with a black coating. The pulse is weak, and the temperature sub-normal. The body weight falls rapidly to much below the normal standard in spite of an appetite which is often voracious. Thirst is excessive and micturition is so frequent that broken nights add to the misery of his condition. The quantity of urine varies from 100 to 300 or more ounces in twenty-four hours and the quantity of sugar ranges from 8 to 10 per cent.

In the *chronic* type the patient is elderly, often florid and well-nourished without any special appearance of ill health, but he complains of unaccustomed weakness, thirst, frequency of micturition and some loss of flesh. Many of these patients have been very stout, so that the wasting is not so apparent. The quantity of urine varies from 60 to 120 ounces, and the quantity of sugar from 3 to 10 per cent.

Prout thought that diabetics were usually fair-skinned and red-haired, and Fagge has confirmed this impression. Such an opinion is of little value, as no one would pretend that all fair-skinned persons are diabetic or that all dark-skinned persons enjoy an immunity from the disease. Babington believed he could always detect diabetics by their odour. We now know that the breath of diabetics contains acetone and ethylic alcohol at times, but many diabetics do not smell at all differently to healthy persons. The *skin* of diabetics is usually harsh, and the hair dry. The face is pale, often cyanosed, and the lips parched. The nutrition of the *nails* is often interfered with; they become brittle and in some cases come off. The *temperature* is usually normal or subnormal, but Skerritt has described a case of *acute febrile glycosuria* in which the temperature ranged for three days between 99° and 103°. The intellectual centres are as a rule unaffected; but there is often *irritability of temper* and other evidence of enfeeblement of the controlling centres; common sensation is usually unaffected; motor paralysis is rare, but sometimes occurs as a consequence of diabetic neuritis. The *knee jerks* are frequently absent, especially in grave cases. *Sexual power* is gradually lost; as a rule this loss is permanent, but if the disease improves it may be regained. In females *menstruation* is as a rule deficient or absent.

Vision is frequently affected, the most common derangement being due simply to weakening of the muscle of accommodation, but toxic amblyopia, retinitis, hæmorrhages, and cataract may occur. *Hearing* is usually normal, though in one of my cases deafness came on at the same time as the disease, and aural complications are not uncommon. *Smell* and *taste* may be blunted. The *appetite* is usually good, and digestion easy, but constipation is very commonly present. The *tongue* is often red and sticky, and the patient usually complains of dryness of the mouth, which sometimes causes a difficulty in deglutition. The *teeth* are generally decayed and falling out, from atrophy of the gums. The abdomen is often retracted, but may present no unusual appearance, and the outline of the liver and spleen are commonly normal, though *enlargement of the liver* is a tolerably frequent occurrence.

The *stools* are peculiarly fœtid; this was first pointed out by Hodgkin who ascribed it to deficient bile.

In uncomplicated cases there is no cough or dyspnœa, and the position and outline of the heart are normal. The *cardiac impulse* is often feeble, and evidences of weak circulation are only too common, especially as the disease advances. *Endocarditis* occurs rarely, but *fatty degeneration* of the muscular wall of the heart is a frequent change.

In early cases, before the heart's wall has degenerated, the *pulse* preserves a good degree of tension, as illustrated in (*Fig.* 12), which shews a full tidal wave. But later on, the heart fails, the pulse fails too, and we get such a tracing as the following (*Fig.* 13).

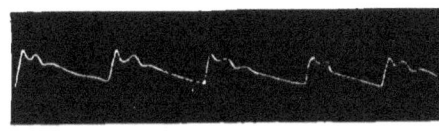
Fig. 12.

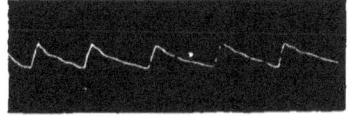
Fig. 13.

The pulse rate is not increased unless in consequence of some complication. It is usually regular, without intermissions. A rapid pulse is one of the early signs of the onset of coma.

The peculiar *odour* of the breath has been already alluded to and ascribed to the presence of acetone and ethylic ether. *Lung* affections, especially chronic pneumonic phthisis, occur very commonly. The symptoms are very latent, cough being slight and expectoration often absent. But such lung troubles are of the nature of complications and will be fully described in that section.

The *blood* contains excess of sugar. Rollo shewed that the blood of diabetics obtained by venesection remained for days without undergoing putrefaction, the sugar acting as an antiseptic. The quantity found has varied in the case of different observers. According to Seegen in mild cases it does not exceed the normal, but in severe examples it may rise as high as 0·4 per cent. Sugar has also been found in the sweat, tears, and saliva.

Very remarkable *fluctuations of body weight* occur during the course of this disease, a variation of two or three pounds being observable in the course of a few days. A patient wrote to me recently, sending the following table of weights, and complaining that her medical attendant scarcely seemed to believe in its accuracy. As this experience has occurred to me many times in hospital practice, I do not hesitate to publish it as illustrating a fact.

June 30th	8 st. 13 lbs.
July 7th	9 st. 1½ lbs.
July 14th	9 st. ¼ lb.
July 22nd	9 st. ½ lb.
July 29th	9 st. ½ lb.
August 9th	8 st. 12 lbs.
August 14th	8 st. 10 lbs.
August 21st	8 st. 12 lbs.
August 28th	9 st. 0 lbs.
September 5th	9 st. ¼ lb.
September 15th	8 st. 11½ lbs.
September 18th	9 st. ¾ lbs.

Between the 15th and 18th of September there was a gain of 3¼ lbs. These variations are probably to be attributed to the very great daily loss in water and solids, amounting to many pounds, by which the normal equilibrium of nutrition is deranged.

The Urine.—The capital symptom of diabetes mellitus is the presence of sugar in the urine, and the diagnosis of the disease depends upon its detection.

Is sugar ever present in quantity in the urine of healthy persons? Physiologists tell us that minute traces can be found in normal urine, but the amount is far too small to be detected in the ordinary way.

Flint has asserted that glycosuria, in obvious degrees, is present in one out of three hundred and seventy-seven healthy persons. This statement lacks confirmation, and is not borne out by my experience. The only healthy, or apparently healthy person in whom I have met with glycosuria was the following:—

CASE 11.—*Glycosuria—no polyuria—no thirst—no loss of weight.*

Arthur M., thirty-two, metal roller, came to me in July, 1887, to be examined for life insurance. He was a very powerfully built man, 5 ft. 6 in. in height, and weighing twelve stone. There was no history of injury, or syphilis, or any family history of diabetes. His knee jerks were present. There was no other evidence of disease but his urine, which was of sp. gr. 1029, and was loaded with sugar. The quantity was not increased. A second sample, examined a fortnight later, was just the same. He was sent again to me in May, 1888, when I found no loss of weight had taken place, and no thirst or polyuria had come on, but the urine was still loaded with sugar, sp. gr. 1023. In order to be sure that there was no fallacy, the urine was fermented and the presence of sugar placed beyond doubt.

Eating excess of sugar on an empty stomach is said to cause temporary glycosuria in healthy persons (Fowler).

In disease, apart from diabetes, sugar is occasionally found in the urine, but not so often as is sometimes asserted. Thus T. A. Mc.Bride, in a discussion at the New York Academy of Medicine, stated that it may be found in the urine of incipient phthisis, emphysema, chronic bronchitis, chronic pleurisy, heart disease, liver troubles, nervous troubles, cerebral hæmorrhage, fracture of skull, concussion of brain, and in psychoses. He had also met with it in cases presenting the following group of symptoms,—insomnia, neurasthenia, paræsthesia, hemi-anæsthesia, and paresis. It is very doubtful whether these statements are correct, except for nervous disorders. Goodhart has published a number of examples of glycosuria connected with neurasthenia, and the occasional occurrence of this symptom in organic disease of the nervous system and in insanity and epilepsy is sufficiently attested to be indisputable.

The *quantity* of urine in diabetes is generally greatly increased, from 200 to 300 ounces in twenty-four hours being a common amount, but in certain cases, especially in elderly persons, the quantity may not exceed 100 ounces, and where diarrhœa is present may not exceed or may even fall below the normal.

The *specific gravity* is usually high, from 1030 to 1050, but again in elderly persons it may be below 1030, and even below 1020. A low specific gravity is no proof of the absence of sugar, nor is a high specific gravity any proof of its presence. The *colour* is usually pale greenish yellow, or straw coloured, but gouty diabetics often pass high coloured urine, loaded with lithates. Diabetic urine is generally *clear*, but may be turbid from the growth of torulæ or from lithates.

The *reaction* is almost invariably strongly *acid*.

It deposits very commonly a considerable amount of uric acid, even in cases in which there is no gout. This is often due to the use of nitrogenous diet. Under the microscope uric acid crystals and torulæ may generally be seen.

Of the normal constituents of the urine there is an increase of water, chlorides (Senator), sulphates, creatinin, and phosphates. Butel has found that the phosphates stand to the urea as 1 to 10. The urea is increased, holding an approximate relation to the amount of sugar of 1 to 22 (Harrison and Slater). The amount of ammonia is greatly increased (Hallervorden). Stadelmann attributes this to the enormous excess of acid in the blood, which in some way disturbs the normal mechanism for fixing ammonia. According to Weil nitric acid is formed normally in the body by synthesis of ammonia, and it is this process which is destroyed, leading to a diminution of nitrates and excess of ammonia.

Of abnormal substances *sugar* is present in quantities varying from 1 to 10 per cent. It has been suggested that where vesical catarrh is present the sugar may be destroyed by a process of fermentation in the bladder, and F. Müller has related a case of pneumaturia in which the bladder contained a mixture of carbonic acid, hydrogen, and nitrogen, derived, he believes, from fermentation of sugar which was present in the urine; but, as Teschemacher points out, if this were the case alcohol should be formed, and so far this has not been demonstrated. Glycogen is said to be constantly present (Leube), and lævulose in some cases (Seegen).

In some cases the sugar may be present intermittingly, as in the following curious case :—

CASE 12.—*Diabetes Mellitus—thirst—polyuria—intermittent glycosuria—rheumatic pains—lichen scrophulosorum—scaly eruption on scalp—amblyopia.*

George B., aged twenty-eight, was first admitted on July 24th, 1886, complaining of lumbar pains, thirst, and polyuria.

History.—The last two symptoms had existed for three months, but the pain for only three weeks. At Christmas he had *rheumatic* pains in his feet and knees, but was not confined to bed, and had never been well since. He could remember no other illness, or any injury. His family history was good, and he knew of no other case of diabetes among his relatives. He had been under treatment by Dr. Foxwell as an out-patient for three months before his admission. His urine averaged over 200 oz., sp. gr. 1040; urea 0·9 per cent.; sugar, 8·0 per cent.; He had lost 14 lbs. in weight. He was sent to the Suburban Hospital and afterwards re-admitted from there on September 26th of the same year. He then complained chiefly of weakness.

Condition on Admission.—He was a fairly well-built and well-nourished man, with a pale and anxious face. He weighed 8 stone 9½ lbs. There was slight ptosis of the right eyelid which had always existed. His skin and joints were normal. Patellar reflexes present. Tongue clean. Bowels confined. No abnormal physical signs. Pulse, 66; Temp., 98; Resp., 24. On ordinary diet he passed over 100 oz. of urine, but on being put on diabetic diet the quantity fell to 50 to 60 oz., while the sugar fell from 8 per cent. to 1·1 per cent. There was generally a faint haze of albumen present.

Progress of Case.—He was treated with Clemens's solution, and under this treatment the urine kept low, the sugar was never more than 4 per cent., often much less, and he gained weight. On his discharge, November 2nd, he weighed 8 stone 12¼ lbs. He was ordered to attend as an out-patient.

After leaving the Hospital he continued to attend as an out-patient, continuing his diet as well as he could, and going on with cod liver oil and iron in the shape of the sulphate.

Nov. 23rd. Urine, sp. gr. 1026; no sugar.

Dec. 7th. Urine, sp. gr. 1017; no sugar. Is eating brown bread; ordered to eat potatoes in moderation.

Jan. 18th. Urine, sp. gr. 1020; no sugar.

Jan. 25th. Urine, sp. gr. 1016; no sugar.

Feb. 22nd. Urine, sp. gr. 1030; loaded with sugar.

March 22nd. Urine, sp. gr. 1025; a trace of sugar.

May 17th. Urine, sp. gr. 1026; no sugar. Is eating no potatoes, but brown bread.

May 24th. Urine, sp. gr. 1027 ; loaded with sugar.
June 9th. Urine, sp. gr. 1027 ; no sugar.
June 28th. Urine, sp. gr. 1028 ; no sugar.
July 26th. Urine, sp. gr. 1018 ; no sugar.
Aug. 23rd. Urine, sp. gr. 1032 ; a trace of sugar.

From this time he continued to have a trace of sugar in his urine up till January 3rd, 1888, but it was only absent on that one occasion. He was seen about every fortnight during all this time. His general condition remained unsatisfactory. He had lost almost 5 lbs. in about eighteen months, and looked pale and worn. When examined in October his liver dulness was found to extend down to the level of the umbilicus.

The next occasion on which sugar was absent was in August, 1888. He had been drinking bitter beer for the last six months.

On November 13th the urine had a specific gravity of 1035, but contained so little sugar that it was analysed, and 4·5 per cent. of urea was found to be present. He complained about this time of bleeding from the gums, which was stopped by a krameria mouth wash.

On December 4th he had an eruption of lichen scrophulosorum on the chest, which was soon cured by ung. hydrarg. ammon.

Dec. 18th. Urine, sp. gr. 1020 ; no sugar.

Jan. 1st, 1889. His urine contained sugar, and continued to do so in varying amounts. In February he complained of a dry, brown scaly eruption on his scalp, which was greatly improved by washing with borax lotion. He began to complain of dimness of vision, and on May 28th he was sent to the Eye Hospital for a complete report upon his vision, which was as follows:—

V. in R. eye is $\frac{5}{60}$ when corrected for error of refraction ; V. in L. eye is $\frac{6}{60}$ badly. Both fundi are normal, and media clear. He smokes over 4 oz. of black twist tobacco a week, and has central scotoma for red and green, the defect arising no doubt from toxic amblyopia. Sees better in twilight, etc.

Albumen is occasionally met with in the urine of elderly diabetics, whose kidneys are often affected by chronic nephritis. It is less common in young subjects. Maguire says he has always found it in the urine of cases of diabetic coma of Küssmaul's type. This is generally true, but there are exceptions.

Indican and skatoxylsulphuric acid may be present in excess (Otto).

Acetone, aceto-acetic acid, beta-oxybutyric acid, and beta-crotonic acid have all been found, and stand in close chemical relation to one another. Much interest attaches to these bodies, as they are believed to be the poisons which give rise to Küssmaul's coma, and some of them give the well-known Burgundy red coloration with ferric chloride. This reaction is also occasioned by formic acid, which, according to Le Nobel, is also present in diabetic urine. Oxalic acid and hippuric acid are also frequently present (Czapek).

Tests for Sugar.—The most certain and the most delicate test for sugar in the urine is fermentation. According to Max Einhorn, it will detect 1-10th per cent., while if the urine is previously boiled for ten minutes it will detect 1-20th per cent. Gerard, of Liverpool, was the first to use this means of detecting glycosuria. He mentions it in a communication which Rollo has published in his book.

The simplest ready means of testing for sugar is Fehling's solution, prepared according to the following formula :—

Sulphate of Copper	90½ grs.
Neutral Tartrate of Potash	364 ,,
Solution of Caustic Soda	f℥iv.
Add water to make up exactly six fluid ounces.	

Two hundred grains of this solution are exactly decomposed by one grain of sugar (Roberts). About a drachm of this solution should be boiled in a test tube, an equal quantity of urine added, and the mixture boiled again. If sugar is present the yellow suboxide of copper is thrown down. Uric acid, kreatinin, glycuronic acid, aldehyde, lactic acid, lactates and lactose are all stated to reduce copper. Munk recommends the addition of 3 to 5 drops of a 15 per cent. solution of calcium chloride to get rid of these substances, but in doubtful cases it is better to ferment the urine. The *quantitative estimation* of sugar is made with the same solution. The necessary apparatus consists of two graduated burettes, a glass flask or white porcelain dish, and a spirit lamp. Measure off two hundred grains of the solution by means of one of the burettes, run it into the flask or dish, dilute it with two volumes of distilled water, and set it on to boil. Dilute the urine to ten volumes with distilled water, fill with it the other burette, and allow it to run drop by drop into the boiling solution until the blue colour of the latter has entirely disappeared. The fluid should be agitated by stirring or shaking constantly, and from time to time the lamp should be removed, and the reduced copper allowed to settle, so that the colour of the fluid can be distinctly seen. The quantity of urine required to decolorize this amount of solution contains exactly one grain of diabetic sugar, whence it is very easy to calculate the amount present per cent. or per ounce. Let us suppose, for example, it required 150 grains of diluted urine, which represent 15 grains of undiluted urine, then $\frac{100}{15} = 6\cdot6$ grains per cent.

A very convenient and inexpensive apparatus for the quantitative estimation of sugar by Fehling's solution is the one known as Gerrard's Glycosometer (*Fig.* 14).

Method.—Take 5 volumes of urine and dilute with water to 100 volumes. Mix well. Fill both burettes to the zero lines with the diluted urine. Put 10 ccs.

(2½ drachms) of Fehling's solution into the porcelain dish and boil. Whilst boiling run the urine first from the *small* burette into the dish in a slow stream until the blue colour has gone from the Fehling's solution. If the contents of the first burette do not suffice, continue from the second till the blue colour is gone. The level of the remaining urine in the burette shews the percentage of sugar. The instrument is graduated to read percentages of sugar between 10 and 1, but should the urine contain more than 10 per cent., then dilute 5 volumes of urine with 200 of water, proceed as before and multiply the result by 2. If there is less than 1 per cent., dilute 1 volume of urine with one of water and divide the result by 10.

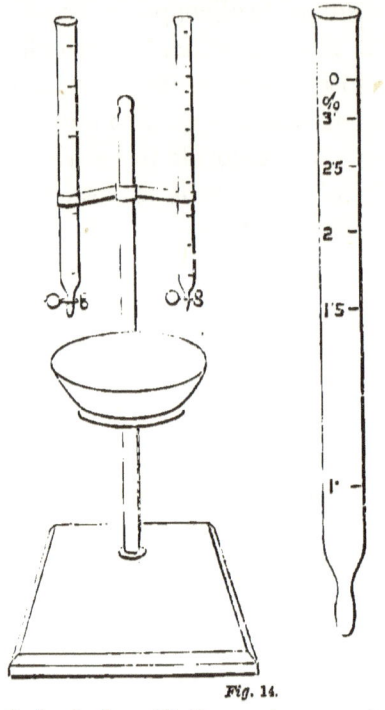

Fig. 14.

The *polarimeter* is a ready method of estimating the amount of sugar, but does not give results that can be relied upon.

There are other tests such as Moore's by boiling the urine with liq. potassæ, when a deep brown coloration appears where sugar is present. The liq. potassæ should always be boiled well before the urine is added, as it sometimes turns brown without the addition of anything. Johnson's, by boiling the urine with a saturated solution of picric acid and liq. potassæ, when, if as much as one grain of sugar per ounce be present the liquid will become quite opaque. Many other substances have been introduced as tests for sugar, but none has shewn itself equal to Fehling's solution, and a perfect method is still a desideratum.

Tests for Acetone.—The best test is that devised by Le Nobel. Pour about an ounce of urine into a urine glass; add a drachm or two of a solution of nitro-prusside of sodium (5 grains to 1 ounce), and a few drops of strong liq. ammonia. After standing a few minutes a rose violet colour is developed, which if much acetone is present may require diluting with water in order to bring out the brilliancy of its colour. After standing some time the colour changes to straw-yellow. On boiling with the addition of acid the colour changes to greenish blue. Ralfe's test is the following:—Boil a drachm of liq. potassæ containing 20 grains of iodide of potassium, and float a drachm of urine on the surface of this solution. At the point of contact a ring of phosphates is formed, which, if acetone is present, becomes studded with yellow points of iodoform; but the objection to this test is that alcohol gives the same reaction, and it may be obtained with lactic acid and other substances. There are several other tests, but the one first mentioned is quite sufficient.

The Ferric Chloride Reaction.—When tincture or solution of the perchloride of iron is added to diabetic urine a rich Burgundy red colour is often developed. Sometimes the colour is obscured by the precipitation of phosphate of iron. It was at one time thought to be a test for acetone, but this is a mistake. It occurs in the presence of aceto-acetic acid, but disappears on heating, or its development is prevented by previously heating the urine. According to Le Nobel β-oxybutyric acid, acetic acid, formic acid and sulphocyanogen compounds, give the same reaction, but in their case it is not affected by heat.

PROGNOSIS.—As a general rule the prognosis of diabetes depends upon the age of the patient. Under forty it is generally a fatal, and a rapidly fatal disease. Over forty it is chronic, often intermittent, and sometimes curable.

But in considering this very important question the etiology of the disease should always be carefully studied and duly considered. In Harker's case of a child of two years of age, who recovered, the disease was apparently due to excessive use of sugar with his food. A family tendency to diabetes, or to nervous affections, is a bad prognostic element. On the other hand, many recoveries have taken place where the disease has supervened after some acute disorder, such as diphtheria. Probably the most favourable cases are those which are distinctly related to gout, especially where the glycosuria is intermittent, and the quantity of urine not very large. Climacteric diabetes is also held to be eminently curable.

The third consideration that must always guide our prognosis is the progress already made by the disease, especially the degree of emaciation caused by it, the quantity of urine, and the total quantity (not the percentage) of sugar excreted.

The fourth point is the influence that dietetic restrictions are found to possess in controlling the excretion of sugar.

But this disease under no circumstances can be regarded as free from danger. The elderly diabetic, whose symptoms can be easily controlled, is liable to die suddenly from some slight imprudence, and is never out of risk. Schmitz has stated that according to his experience those cases in which spontaneous sweating occurred were always attended by mild general symptoms.

DURATION.—It is true, as a general statement, that diabetes runs a longer course in direct proportion to age. In other words, it is an acute and rapidly fatal disease in children and young persons, lasting only weeks or months, or perhaps a year or more. Nevertheless, I have met with cases in children that seemed to run a very chronic course; and Pavy has related instances lasting five, eight, and even more years.

CASE 13.—*Diabetes Mellitus—thirst—polyuria—wasting —three members of family affected.*

Mary Ann L., aged ten, (sister to Agnes L., Case 28, page 125), *had been ill four or five years*. Another sister had died of diabetes, aged eleven-and-half. She was poorly nourished, weighed 2 st. 7¾ lbs. Urine 124 ounces, sp. gr. 1025, acid, straw colour, urea 0·8 per cent.; no albumen; sugar 3·9 per cent. After admission she remained in the hospital five months, gaining weight and improving somewhat on anti-diabetic diet, but she was difficult to feed. There were no incidents worthy of notice during her stay, nor any other points of interest in her case.

On the other hand, in elderly persons the disease lasts for years. Worms has given instances where the disease extended from twelve to twenty-five years, and disappearance of the symptoms is not uncommon.

Termination.—Death occurs in many cases as a result of one or other of the complications to be described, of which pneumonia and pneumonic-phthisis are the most common. In very many cases the patient's strength is gradually reduced, and he dies quietly in a drowsy state without being actually comatose. But in a large proportion of cases death is more or less sudden, the fatal symptoms supervening as the result of some slight occasioning cause, such as a walk too great for their enfeebled powers, a chill, excitement, etc. In these cases death is preceded by coma. This mode of death is so peculiar that it has attracted much attention, and will be described in a separate chapter.

COMPLICATIONS.

INTEGUMENTARY SYSTEM.—The skin of diabetics is generally dry and harsh, with not uncommonly a tendency to desquamation of the epidermis. The circulation is generally feeble, and there is often some cyanosis of the ears, nose, and cheeks, while the feet and legs are cold. *Acne pustules, boils,* and *carbuncles* occur not unfrequently in elderly diabetics. The most common affection of the skin is a form of *erythema* on the hands and arms, legs and feet. The patches are slightly raised, irregularly oval in shape, about half to three quarters of an inch in their longest diameter, and purple in colour.

Diabetes Mellitus—very slight polyuria—thirst—hunger —erythematous eruption on extremities—cataracts—deafness—sciatica—diarrhœa.

CASE 14.—Elizabeth R., aged thirty-seven, shopkeeper, was admitted into the General Hospital on May 30th, 1889, complaining of weakness of the arms and legs, thirst, and polyuria, with pain in the back and abdomen after food, and an eruption on the nape of the neck and on the feet ; also of failing sight. She had been quite well up to four years ago, when, being pregnant for the second time, she suffered from thirst and swelling of the legs, which continued after her confinement, and she became weaker and thinner. She kept a mangle and a little shop up to four months ago, when her parents took her home to live with them.

Family History.—Her own mother died of phthisis, but she had a step-mother. Her father was alive and in good health, but very nervous. Two brothers and a sister were alive and well. Patient was a widow, her husband having died of bronchitis two years ago. She had had two children, both dying after a few weeks of life.

State on Admission.—She was fairly developed, but greatly emaciated, skin dry, some patches of eczema, with enlarged lymphatic glands in the nape of the neck, due to *pediculi capitis*. Skin of lower part of legs and feet roughened and mottled with irregular discoloured areas of congestion, varying in size from a sixpence to a shilling, which itched when warm. These had been coming out in successive crops for eighteen months. The spots when fresh were raised above the general surface of the skin. Temp., 97·6°; Pulse, 96; Resp., 20. Weight, 5 st. 8½ lbs.

Alimentary System.—For three months hunger and thirst had been greatly increased, but she had been taking ordinary plain diet till a few days ago, since which she had taken less bread and potatoes. Tongue furred; teeth all gone but four; a bitter taste in the mouth. Complained of a gnawing pain after food, and of flatulence. Bowels open. Liver dulness in V.M.L. five inches; splenic dulness in M. A. L. one and a half inches; abdomen flat.

Circulatory System.—Complained of shortness of breath and cough. Heart's apex in 5th I. S., inside V. M. L.; sounds clear. Pulse small, regular, compressible.

Respiratory System.—Cough, with scanty mucous expectoration; percussion note resonant everywhere over lungs. Breath sounds normal except below the angle of the left scapula, where there was a patch of bronchial breathing and crepitation after cough. Voice sounds not altered.

Nervous System.—Complained of frontal headache, with giddiness at times; also of pain at the bottom of the back; but had no neuralgic pains or abnormal sensations. The left ear had been deaf for nine months, and the eyesight had been failing for eighteen months. There were cataracts in both eyes, and the fundi could not be seen. Knee jerks absent.

Genito-urinary System.—She had not been regular for four years, but there was some yellow vaginal discharge. She thought the quantity of her water was about eight pints; and a specimen examined was clear, acid, 1026, pale brownish-yellow, loaded with sugar, and containing a faint cloud of albumen.

On June 1st, three days after admission, the urine report was as follows:—68 oz., 1020, acid, pale opalescent;

urea, 0·9 per cent.; sugar, 2 per cent.; albumen a very faint haze, blood in traces. Under the microscope, red and white blood corpuscles and squamous epithelium. The blood and albumen were thought to be due to the vaginal discharge. This was on milk diet with bread.

Treatment.—She was then put upon the following diet: —Meat, green vegetables, a pint of beef tea, Bonthron's gluten bread, Vichy water and lemon juice. She was ordered, in addition, a teaspoonful of cod liver oil three times a day.

June 4th. She complained of severe pain shooting down the right arm and leg; it shot down the back of the leg from the waist to the foot. For these neuralgic pains she was ordered *Sodii Salicylatis*, gr. x., in water thrice daily, with benefit. She could not manage her diet, which was modified by the substitution of fish, chicken, and meat jelly for meat, and the addition of half a pint of milk.

June 21st. She had an attack of diarrhœa, which was treated with *Bismuth Mixture*. The cod liver oil was discontinued, and the diarrhœa, which was apparently caused by it, ceased.

June 29th. As the pain still continued she was ordered *Antipyrin*, gr. x., three doses to be given at hourly intervals, when the pain was severe, but without much advantage. She was ordered to resume the *Salicylate of Soda* and the cod liver oil.

July 4th. Ordered to stop other medicines and to take *Antipyrin*, gr. xv., three times a day.

July 8th. The diet was again changed to mutton, potatoes, beef jelly, and milks, and 6 oz. of wheaten bread, with Vichy water and lemon juice.

July 16th. For the last three days there had been slight swelling of the eyelids with watery discharge, and there was some conjunctivitis with a small sub-conjunctival hæmorrhage in the right eye.

This morning an extensive eruption broke out over the forearm and hands. It consisted of rose-coloured elevated spots, as large as a split pea, thickly distributed over the elbows, fingers, and palms of the hands, and often coalescing so as to produce red shiny blotches. Some of the larger spots had a tendency to become white and transparent in their centres, looking something like nettle stings. This papular erythema gradually faded until by the 22nd there was only a brownish-red mottling over the seat of it. Shortly after this she was made an out-patient (July 24th).

The following table shews the state of the urine, week by week, and the weight of the patient.

DATE.	URINE	UREA.	SUGAR.	WEIGHT OF PATIENT.
June 10th.	20 oz.	1·7 p. c.	1·3 p. c.	5 st. 10¼ lbs.
,, 19th.	24 oz.	1·7 p. c.	0·4 p. c.	5 st. 8¾ lbs.
,, 24th.	42 oz.	4·7 p. c.	1·04 p. c.	5 st. 8¾ lbs.
July 3rd.	28 oz.	2·5 p. c.	—	5 st. 7¼ lbs.
,, 9th.	36 oz.	2·3 p. c.	0·5 p. c.	5 st. 8¼ lbs.
,, 16th.	32 oz.	2·6 p. c.	6·4 p. c.	—
,, 23rd.	56 oz.	1·2 p. c.	6·8 p. c.	5 st. 8½ lbs.

This shews that the sugar could be controlled by diet, but without any benefit to the patient's general condition. She grumbled so at the necessary restrictions that she was allowed to resume ordinary diet with consequent return of sugar, before her discharge.

Leroux has described the case of a child in whom a symmetrical erythema in large elevated itching patches attacked the nose and chin. This was followed by an outbreak of psoriasis guttata. Later on the nails underwent a peculiar change, which led to their falling off, and finally, after copious sweating, a lichen-like papular eruption appeared, which was followed by numerous boils. Davies Pryce has described a condition which he has called erythematous œdema; he believes it to be dependent upon diabetic neuritis. It is characterised by very severe gnawing pains, with swelling and discoloration of the skin. *Burning* of the palms of the hands and soles of the feet (Marcet) may be complained of, while sweating in the same situations is not uncommon. In fact profuse general sweating sometimes occurs, as in a patient of mine, a male, aged thirty-nine, in whom it was checked by Dover's powder after atropine had been tried and failed.

Eczema of the genitals is somewhat common, especially in women, and may constitute a serious trouble. It is undoubtedly set up by the irritation of fungoid growths in the saccharine fluid remaining on the parts, and in males a very little cleanliness will prevent this. Even in men, however, an eczematous balanitis may occur. In women the difficulty of keeping the parts thoroughly clean is more difficult, and eczema attended by great irritation, which spreads over the abdomen and thighs and renders life a misery, is too often seen. In some cases *pruritus* vulvæ is complained of without any eczema being present. *Purpura* may be present in the prodromal stage (Simon).

Kaposi has described an affection to which he has given the name of *papillomatosis diabetica*. The patient was a Brazilian doctor, who had suffered from diabetes for twenty years, but was well-nourished and vigorous. The affection was limited to the left arm and forearm which were extensively swollen, and the backs of the fingers were covered with excrescences varying in size from a lentil to a sixpence, ulcerated in places, the ulcers being rounded or kidney-shaped, bordered by florid granulations, and discharging freely. On the elbow was a growth as large as the palm of the hand, raised over two centimetres above the level of the surrounding skin, deeply fissured and presenting a slightly bleeding surface covered with warty protuberances.

Addison and Gull first described a peculiar eruption which is now called *xanthoma tuberosum*. A case was described by Dr. Bristowe in 1866. In his case the eruption consisted essentially of somewhat indurated tubercles of a dull reddish hue (but not much deeper in colour than the surrounding skin in their neighbourhood) and of roundish or obtusely conical form. Their margins passed invariably into the healthy skin around, and their apices were often of a pale yellow colour, as though containing a minute quantity of pus. Their size was not uniform; to speak roughly their horizontal diameter varied from a line upwards, and their vertical projection from a line downwards. The yellowness of their apices was found not to be due to any accumulation of fluid there; for this part, like the rest of the tubercles, was quite solid. The tubercles appeared from microscopical examination to consist essentially of a kind of dense fibrillated texture, studded more or less with oil globules of various sizes. It was the presence of such globules in great abundance that caused the yellowness just described.

Marchal (de Calvi) quotes a case, perhaps of this nature, where the body of the patient was literally covered with great coppery pustules or hard vesicles, containing a material as hard as very dry cheese. They were harder and more elevated than variola pustules. Some were rounded, terminating abruptly in a point. Generally isolated, they were occasionally in groups. This eruption disappeared, leaving only minute cicatrices, and did not recur as the disease got worse.

This tendency to disappear is true of xanthoma.

Another peculiar eruption, in the form of circumscribed necrotic patches has been described by Rosenblath. It occurred in the form of small round red spots the size of pins' heads, about the ankles and on the dorsum of the feet. Some of the spots were little vesicles containing watery fluid. At the autopsy they were found to consist of little centres of necrosis, and ulceration was found on the tongue and in the mucous membrane of the œsophagus, stomach, and intestines. The skin eruption appeared to commence in the sweat glands.

Cellulitis and *gangrene* are complications which are more apt to occur in diabetes as life advances, and are more common in men than in women, though Hunt throws doubt on this latter statement.

CASE 15.—*Diabetes Mellitus—gangrene of both lower extremities—successful amputation of necrosed parts.*

Joseph F., sixty-four, tailor, was admitted February 20th, 1889, with gangrene of right foot.

History.—Father died at eighty-six; cause not known. Mother was killed. Three brothers alive; one brother died young. No history of illnesses. About five years ago he came under Mr. Chavasse with what he called "corns" on the second digit of each foot. That of the left foot got well, and the toe on the right foot was amputated. At that time he had no suspicion of any illness. He had not complained of thirst or of weakness; but on coming to the Hospital he was told he had diabetes. For some time he was only slightly troubled with sores on his toes, and about eighteen months ago he began to be thirsty. Twelve months ago last March all the toes on the left foot were taken off in the General Hospital, and he then went to the Suburban Hospital, leaving there in July. In January of this year the toes on the right foot began to be affected, and in February he came into the General Hospital again, and the third and fourth toes were amputated at different times. In April the ankle began to be bad, and on June 15th the leg was amputated. Latterly his health has not been so well. It is only in the last twelve months he has noticed he has been passing a large amount of water. Has had no cough. No history of boils or other skin diseases.

Whilst the urine was measured from April 3rd to 21st he was passing from 66 oz. to 109 oz. in twenty-four hours, with an average of about 90 oz.

Urine Report:—

Date	Quantity	Sp. gr.							
Feb. 25.	—	1036.	Ac.	Urea 1·4 p.c.	Sugar	8·3 p.c.	Acetone none
Mar. 29.	—	1036.	,,	,, ·8 ,,	,,	5.5 ,,	,, present.
April 1.	70 oz.	1034.	,,	,, ·5 ,,	,,	5·2 ,,	,, ,,
,, 6.	88 oz.	1030.	,,	,, — ,,	,,	5·5 ,,	,, ,,
,, 15.	80 oz.	1036.	,,	,, — ,,	,,	5·0 ,,	,, —
June 17.	—	1033.	,,	,, 1·0 ,,	,,	6·2 ,,	,, —

CASE 16.—*Diabetes Mellitus—polyuria—thirst—albuminuria—gangrene of toes—knee jerks diminished—improvement on diet and opium.*

William F., aged fifty-seven, attended as an out-patient on October 26th, 1886, complaining of passing three to four quarts of water and of getting up very frequently at night for this purpose. He had been ill three years, when he got gangrene of three toes on the right foot, and his urine was found to contain sugar. He suffered from thirst; his urine was pale, clear, sp. gr. 1022, and contained a cloud of albumen and a large quantity of sugar. The knee jerks in both legs were almost absent. On anti-diabetic diet and opium he lost his thirst, he was not disturbed at night to make water, and the quantity of this secretion fell to 90 oz. daily, or little more than two quarts, of sp. gr. 1020. He was not seen after January 18th, 1887.

Kaposi has described a rare condition under the name of *gangræna bullosa serpiginosa*. The patient was a woman, aged fifty-one; on her left leg were three gangrenous patches, and over the neighbouring sound skin fifteen or twenty bullæ varying in size from a pea to a bean. The condition improved somewhat under treatment at Carlsbad, then got worse and the patient died.

Several cases of *perforating ulcer* of the foot in diabetes have been published (Kirmissen, Heusoner,

Spencer). The ulcer is preceded by circumscribed anæsthesia, and an anæsthetic zone surrounds the ulcer when formed. In Heusoner's case the ulcer preceded the symptoms of diabetes, and *post mortem* nothing abnormal was found in the medulla or pons. It is probable that this lesion is another consequence of disease of the peripheral nerves.

Against this formidable catalogue of skin affections due to diabetes it is some comfort to set the fact that persons have been known to lose chronic skin affections upon the supervention of diabetes (Watson).

Kaposi attributes all diabetic skin affections to impregnation of the skin with sugar. We have seen that in the case of eczema genitale, the sugar is undoubtedly the exciting cause, but its direct influence is doubtful in other cases. We know that all the tissues of diabetics are very prone to disease in consequence of malnutrition, but it is taking a somewhat narrow view to ascribe everything to the sugar circulating in the blood, which not unfrequently is not more than normal, the overplus being constantly excreted by the kidneys.

Dropsy.—In a disease characterised by polyuria it is hardly to be expected that dropsy should occur. It is, however, sometimes present, evidently as a result of heart failure and is attended by a cessation of the polyuria. Ascites may be present as well as œdema of the lower or even upper extremities (Roberts).

Case 17.—*Diabetes Mellitus—bronchitis—œdema of trunk and lower limbs—recovery.*

Mary Ann M., aged thirty-nine, widow, metal button maker, was admitted into hospital on January 8th, 1887, complaining of cough, pain in the back, and swelling of the legs and abdomen.

History.—She had been ill six weeks, the attack commencing with cough. A week ago she noticed her abdomen was swollen, and soon after the legs began to swell. Her breath got very short, and she felt very ill and weak, but she went about until the day of admission. She had been quite strong, except that ten years ago she had rheumatic fever, and for the last five or six years had suffered from a cough in the winter.

Condition on Admission.—There was much cyanosis of the face, hands, and arms, with great œdema of the lower extremities and trunk, but no ascites or hydrothorax. She could not lie down in bed, and her breath was very short and wheezy. Pulse 114, Temp. 98°, Resp. 30. Troublesome cough; no dulness over lungs; breath sounds very wheezy, with moist râles and sibilant rhonchi throughout; vocal resonance undiminished; heart's action regular, no murmur; pulse very small and weak. She had not menstruated for two months, and the flow had been very scanty for some months. Urine 48 oz., acid, 1043, a faint cloud of albumen, 8·3 per cent. of sugar, no blood, no casts, squamous and pear-shaped epithelium in deposit. She was put on strict antidiabetic diet, with a little morphia for her cough. On this treatment she steadily improved, the sugar diminished rapidly, and the œdema disappeared. She left the hospital on February 21st, free from dropsy and with no sugar in her urine, though this still shewed a faint trace of albumen.

Temperature.—The temperature in diabetes is usually normal or sub-normal, and in some cases may be very low indeed, as in the example given by Fagge, where the thermometer marked only 93·6. In the following case the temperature was irregular, being sometimes as high as 100°, sometimes as low as 95°.

CASE 18.—*Diabetes Mellitus—polyuria—wasting—diarrhœa—very low temperature—discharged unrelieved—subsequent death.*

George H., forty-two, tailor, was admitted into hospital on Jan. 25th, 1887, complaining of tingling of his tongue and the roof of his mouth, with cramps in his legs. These symptoms came on about a week after Christmas.

History.—He had been losing weight since November. He had been passing more water than usual and had been thirsty. He had never been ill, except that when he was nineteen he spat a little blood. Father died aged thirty-five of "brain fever;" mother died aged seventy, of "diseased stomach." There was only one sister who died when she was five, but he did not know the cause of death. He had never met with any accident or personal injury.

Condition on Admission.—He was a poorly nourished man. Pulse 90, Temp. 100°, Resp. 17. Physical examination revealed nothing, except some want of resonance and deficient breathing at the apices of both lungs, and his pulse was very weak. Urine, 116 oz., sp. gr. 1035, acid, straw-coloured, no deposit; 0·8 per cent. of urea, 5·8 per cent. of sugar, no albumen.

Progress of Case.—He was dieted and treated with alkalies and cod liver oil; afterwards with morphia. He had several attacks of diarrhœa which interfered with his progress, and his temperature was irregular, sometimes rising to nearly 100°, but more often being subnormal, being several times as low in the morning as 95°. The low temperature was not the result of the diarrhœa, that is, it did not occur at the same time as the diarrhœa, or after those attacks. He never complained of feeling cold when it was at its lowest, and on being asked said he did not feel cold. He was ultimately discharged, and we heard died in a short time.

In acute febrile diseases the sugar sometimes disappears from the urine, as in a case of enteric fever, reported on p. 125. It has also been observed to disappear in relapsing fever (Simon), small-pox (Rayer, Pavy), febrile angina, dysentery (Andral), and pneumonia (Leube, Oliver).

NERVOUS SYSTEM.—The connection between diabetes and grave diseases of the nervous system has been fully illustrated in the section on etiology, it is, therefore, not surprising if in this disease we meet with listlessness and depression of spirits, weakness and peevishness of temper (Watson), or even melancholia with suicidal tendencies (Legrand du Saulle), or temporary mania (Pavy). In certain cases there seems to be an alternation between the mental disturbance and the glycosuria, the latter only appearing when the patient's mental state is relatively good (Madigan).

In some cases there may be symptoms resembling those due to an intra-cranial growth, *e.g.*, headache and giddiness, with epileptic or apoplectic attacks.

Diabetic *neuralgia* is usually symmetrical, though it may commence on one side. It usually comes on suddenly, frequently in bed, and the pain is excruciating. Each attack lasts only a short time, but there may be three or four in twenty-four hours. The attacks appear to be aggravated by the warmth of the bed. The sciatic nerves are specially liable to be attacked (Cornillon).

CASE 19.—*Bilateral Sciatica—glycosuria—albuminuria —no polyuria or thirst.*

Ebenezer H., sixty-five, came as an out-patient on May 14th, 1889, complaining of paroxysms of excruciating pain in the course of the sciatic nerves.

History.—He said the attacks lasted six to eight hours, and that they had come on about once in three months for the last twenty years. For thirty years he had been subject to attacks of gout. He did not know of any case of diabetes in his family.

Condition on Admission.—He was a short, stout, florid man ; he had lost most of his teeth ; his bowels were confined. Physical signs normal. Urine, sp. gr. 1013, acid, a cloud of albumen ; reduced Fehling moderately. He was dieted, and ordered to take the following :—

℞ *Sodii Salicylatis* gr. xv. ; *Infusi Gentianæ* ʒj. ; to be taken thrice daily.

May 28th.—The urine was free from sugar, and there had been no more paroxysmal pain.

CASE 20.—*Diabetes Mellitus—thirst—polyuria—wasting —unilateral sciatica.*

Josiah W., fifty-six, came to the out-patient department on April 1st, 1886, complaining of pain and swelling in the right leg, which he had been told was sciatica. He also complained of violent pain in the abdomen, of great thirst, and of passing a large quantity of water, as much as three pints in the night alone.

History.—He had been under treatment for six weeks, but the pain began a month earlier. He had been losing weight. He had usually enjoyed good health, but eighteen years previously he had "rheumatics," and was laid up five weeks, with pains which shifted about; his knees were swollen at the time. He had bronchitis twelve years ago; he had never had gout. Had a comfortable home; said he was temperate, but drank two or three pints of beer daily. He used lead in his work to hold castings, but there was no lead dust. Father died, aged sixty-three, of apoplexy. Mother died, aged sixty-three, of dropsy. Of his thirteen brothers and sisters five were living in good health; the others died in infancy, except one sister who died in childbirth.

Condition on Admission.—He was a big, stout man, with a florid face; weighed 14 st. 3 lbs. No blue line on gums. Right leg pitted on pressure, but there was no obvious œdema, except some puffing over the instep. No œdema elsewhere. Pulse 88, Resp. 18, Temp. 98·8°. Physical signs normal. Breath foul; tongue red at edges; suffered from wind; bowels regular. Urine 76 oz., loaded with sugar, a cloud of albumen.

Progress of Case.—He was put on diabetic diet with an alkaline mixture, and the sugar at once disappeared. The pain rapidly improved. On being allowed ordinary diet the urine contained 2 per cent. of sugar, which disappeared again at once on the diabetic diet being resumed. The swelling and pain completely disappeared, and before he left he was allowed a little brown bread, without giving rise to any return of the glycosuria. He was discharged on May 5th, and made an out-patient.

Florain has described a painful affection of the fingers, like the *pricking* of a pin, which occurred in a pregnant woman whose urine was loaded with sugar. Intense hyperæsthesia of the soles of the feet has been described by Auerbach.

Loss of sexual desire is an early and very frequent symptom.

Paralysis may attack single groups of muscles, giving rise to *ptosis, strabismus,* or *paralysis* of a limb. This is probably not due to central disease, but to peripheral neuritis, examples of which have been published by Althaus and Buzzard. In other cases there may be complete *paraplegia,* affecting both upper and lower limbs, mobility, sensibility, and reflexes being all abolished, without paralysis of the sphincters (Strahan). This form is also evidently due to general peripheral neuritis. In some cases there may be symptoms closely resembling *locomotor ataxy,* though the pupil reflex is said to be never lost (Fischer).

Salomonson has described a case in which, after violent exercise in skating, the patient complained of pains in the body and legs. Sensibility and gait were normal, and there was no loss of co-ordination, but the superficial reflexes were impaired, and the patellar reflexes on both sides were abolished. She could stand with her eyes shut and the pupils reacted well to light. The right eye was healthy; the left shewed some opacity of the lens. There was great diminution of the galvanic excitability of the muscles and nerves. He regarded it as a case of peripheral neuritis due to diabetes.

Lépine and Blanc have described a case of hemiplegia with only microscopic lesions in the motor convolutions.

Apoplexy occurs sometimes, but is not common.

CASE 21.—*Diabetes Mellitus—no polyuria—wasting—vertigo—albuminuria—apoplexy—death.*

Mr. W., fifty-six, was seen in 1885, as a candidate for life assurance, when sugar was found in his urine. He was brought to me on November 29th, 1886, by Mr. F. W. Underhill, of Moseley, complaining of giddiness and temporary attacks of loss of consciousness. His eyesight was good; there was no hemiopia, or double vision; his optic discs were normal. There was no obvious hypertrophy of the heart; the pulse was rather empty. The urine was not increased in amount; he did not rise at night to pass it; it was acid, sp. gr. 1021; contained a little sugar, and gave a haze of albumen on boiling.

I heard that he got much better on the treatment employed; but on Feb. 3rd, 1887, he had some friends to dinner and drank champagne. The following day he was seized with deafness, inability to speak, and progressive stupor. When seen late that night he could not speak, his pupils were contracted, his pulse hard, and he was restless; but there was no paralysis of the limbs. He was bled to 16 oz., and afterwards expressed himself by signs as feeling better. However, he died the following night from œdema of the lungs.

Coma occurs frequently as the termination of diabetes, and will be described separately.

Epileptic Fits occur occasionally.

Finlayson has described a case in which repeated fits alternating with furious mania occurred for seventeen hours before death. The urine was quite free from albumen and on *post mortem* examination there was nothing to account for the convulsions.

Loss of the Knee Jerk has been observed very commonly in diabetics; it is attributed by Strumpell to degeneration of the peripheral nerves. Dreyfous has met with an example of exaggeration of the knee jerk in a diabetic woman aged sixty-eight. According to Bouchard it was absent in nineteen out of sixty-six cases, or in 29 per cent. Of those in which it was absent 6 or 31·5 per cent. died, while of the others only 2 or 4·2 per cent. died, which suggests that absence is of very bad prognostic significance. Still it is known that the jerk may reappear if the symptoms improve (Marie and Guinon). Rosenstein does not agree in regarding the loss as of grave import, as he found that it bears no relation to the amount of sugar or to the acetone or ferric chloride reaction-giving substances in the urine. He points out that it cannot be reproduced by subcutaneous injections of strychnine, as in alcoholism.

Barnes has described a case in which the symptoms of diabetes and *exophthalmic goitre* came on simultaneously. The patient was a domestic servant, aged thirty-four; she died after being under observation about three months, but unfortunately no examination of the body could be obtained.

EYE AFFECTIONS.—The following table of one hundred and forty-four cases, collected by Galezowski, shews the relative frequency of various eye affections in diabetes.

DISEASES.	CASES.	PERCENTAGE.
Conjunctivitis and Disturbance of accommodation	5	3·5
Keratitis	4	2·8
Iritis	7	4·9
Choroiditis gummatosa	4	2·8
Cataract	46	31·0
Retinitis	27	19·0
Amblyopia	31	21·7
Hemiopia	4	2·8
Paralysis of ocular muscles	10	7·0
Detached retina	3	2·1
Atrophy of optic nerve	3	3·1

He thinks of these, cataract, retinitis, amblyopia, hemiopia, and muscular paralysis, alone, are to be regarded as really dependent upon diabetes; the others are merely accidental complications.

Deutschmann has found in diabetic eyes swelling and softening of the pigment layer on the posterior surface of the iris. In the deepest layers of the lens he found leucocytes containing myelin, with irregular lumps of albuminous material lying between the lens fibres and fine granulation and vacuolation in the peripheral fibres, the epithelium of the anterior capsule stained irregularly, with granular degeneration of the nuclei both at the equator and the periphery. He thinks the changes in the lens depend rather upon the general mal-nutrition than upon the presence of sugar.

Diabetic Cataract is bilateral, developing and ripening quickly; it occurs in younger persons than ordinary cataract, and is attended by severe symptoms of diabetes. It was noticed by Mackenzie and Duncan, but was first clearly described by France in a paper in the Ophthalmic Hospital Reports for 1859. These cataracts are usually soft, but not always (Graefe, Wilde), and they may disappear spontaneously (Nettleship). Operations for their removal are generally undertaken on young subjects, but these are not free from danger, as there have been several cases of death from coma following this operation (Spencer), and the healing process is liable to disturbance by attacks of iritis, etc., (Samelsohn).

Diplopia is mentioned in a case recorded by Willan and quoted by Rollo ; it is due to paralysis of the external rectus on one side. A more common condition is amblyopia, depending upon paralysis of the muscle of accommodation, with loss of converging power, which Trousseau was the first to point out as an early symptom of diabetes. It may be only temporary, passing off after a time (Wallace Anderson).

CASE 22.—*Diabetes Mellitus—brought on by an attack of influenza—amblyopia from failure of accommodative power—pruritus vulvæ.*

O. M. B., aged twenty-two, single, housemaid, was admitted into hospital on April 15th, 1890, complaining of great hunger, thirst, and polyuria. Her illness had lasted about three months.

Family History.—Mother healthy, father asthmatic; two brothers healthy, one delicate; one died of some brain disease, aged eighteen, and another died in infancy; a sister healthy. No history of diabetes.

Previous History.—Was never very strong; in a convalescent home some years ago; in service for five years; had suffered from indigestion and constipation. Just before Christmas she went to the Eye Hospital on account of inability to read.

History of Present Illness.—Some time before last Christmas she had an attack which she called influenza; it lasted about a fortnight, and left her very weak. Soon after this she noticed that she was thirsty and had to pass water more frequently, while she began to lose flesh. Towards the end of February she began to have hysterical fits two or three times a day, and there was a vaginal discharge with much itching and irritation of the vulva. For this complaint she attended the outpatient department of the Women's Hospital. She always had a bad appetite till five weeks ago, since which time she had a constant desire to eat.

Present Condition.—Patient was a pale-faced brunette; she said she used to be quite stout, even up to Christmas. She weighed 6 stone 11¾ pounds. No jaundice, œdema, or cyanosis. Pulse 60; Temp. 98; Resp. 24.

Alimentary System.—Many teeth were decayed, but she had not had toothache. Mouth dry; tongue dark red, clean, moist; hunger and thirst were constant. Had some pain in back, and occasional nausea after meals.

Bowels confined for two days. Abdomen hard, retracted, and rather tender. Liver 4 inches in V. M. L. Spleen 1½ inch in M. A. L.

Circulatory System.—No palpitation or dyspnœa. Heart's impulse in 5th I. S. one inch internal to V. M. L. Area of dulness not increased. Sounds rather weak and prolonged. No murmur. Pulse 60, regular, tracing shews distinct evidence of increased tension, though the curve is small. (*See Fig. 15.*)

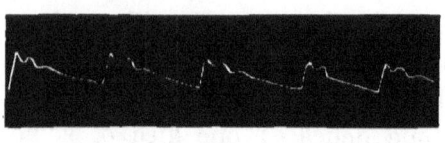

Fig. 15.

Respiratory System.—Normal.

Integumentary System.—There were a few patches of brown pigment on the abdomen near the umbilicus. No eczema.

Urinary System.—She had to rise three or four times at night to make water. Urine 106 oz., pale straw colour, acid, sp. gr. 1042; urea 1·1 per cent.; sugar 10 per cent; no albumen; no blood. There was at times a faint haze of albumen, which was probably due to admixture with vaginal discharge, as it always coincided with the presence of leucocytes and squamous epithelium in the deposit. There was a marked ferric chloride reaction, but no acetone.

Vision.—Her eyes were carefully examined by Dr. Young, who found that vision corrected by glasses equalled $\frac{5}{6}$ in each eye, and that there was no reduction of the field of vision or any ophthalmoscopic change. He reported that her inability to read was "purely due to weakened accommodation and a certain amount of retinal asthenopia arising from her weak state."

All the muscles of the eye may be affected by paralysis. Fienzal has related a case of sudden facial paralysis with corneal ulceration which ended in complete recovery.

There was no history of syphilis or rheumatism, but the urine was loaded with sugar. *Keratitis* and *iritis* are not uncommon.

Leber classifies the *retinal affections* of diabetes in the following manner:—

(1) Simple retinal hæmorrhage;
(2) Retinitis apoplectica;
(3) Retinitis with hæmorrhages and white patches;

To these Hirschman has added:

(4) Retinitis diabetica exudativa.

These retinal changes often closely resemble those met with in Bright's disease. Hæmorrhages are conspicuous, often leading to secondary parenchymatous retinitis (Gowers). White spots are frequently present, and sometimes, though rarely, there may be a peri-macular circle of white patches, but this is never fan-shaped (Samelsohn).

Diabetes Mellitus—polyuria—wasting—pruritus vulvæ—heredity—phthisis—retinal hæmorrhages—acetonuria.

CASE 23.—Amelia M., fifty-seven, housewife, was admitted into the General Hospital on November 15th, 1889, complaining of weakness of sight, pain in the back, loss of strength, and irritation of the pudenda. Her symptoms had been coming on since the climacteric period, which commenced with her four years ago, though there was nothing more definite than weakness and wasting until the last six months. She had had no thirst, and polyuria had occurred only lately. The attacks of pudendal irritation had, however, troubled her from time to time for the past ten years, especially when pregnant.

Previous History.—She had enjoyed a comfortable home and good food, with half a pint of beer daily. She had had erysipelas twice, twenty-four and twelve years ago. She had small-pox at twelve years of age; could recollect no accidents or injuries. Had been married thirty-eight years, and had borne seventeen children besides five miscarriages. Ten of the children were alive and well; the others died young.

Family History.—Father died, aged sixty-four, of bronchitis; mother died, aged seventy-three, of "cancer of the face." One sister had died of phthisis and a brother of diabetes. Another sister died of fever in America. The others (eight) were alive and well. She knew of no cases of gout, rheumatism, or insanity, and of no other cases of diabetes or phthisis.

State on Admission.—She was a well-developed stout woman, with a florid face, moist skin, and great varicose dilatation of the veins of the legs. No œdema. Temp. 98°, Pulse 96, Resp. 24.

Alimentary System.—Appetite poor; no thirst; no discomfort or pain after food. Teeth very defective; tongue clean. Bowels confined; liver dulness in V. M. L., $4\frac{1}{4}$ inches; splenic dulness in M. A. L., 1 inch.

Circulatory System.—Area of cardiac dulness not increased. Heart's apex in fifth I. S. internal to V. M. L., sounds normal. Pulse 96, full, incompressible, regular.

Respiratory System.—Appearance of chest normal. Above right clavicle percussion resonance diminished, vocal resonance increased, expiration prolonged.

Ophthalmoscopic Appearances.—Right eye: Numerous punctiform hæmorrhages round disc, and many glistening white patches, varying in shape, and some of considerable size. Left eye: One small hæmorrhage on the inner side of disc, and in yellow spot region a large area of white glistening patches with numerous punctiform hæmorrhages.

Genito-urinary System.—Had not menstruated for three and a half years. There was no eczema about vulva; the pruritus had existed for three months. Urine 56 oz., sp. gr. 1035, acid, yellow, urea 1·8 per cent, a trace of albumen, sugar 6 per cent., acetone present.

Treatment.—House diet, no sugar ; *Ext. Cascaræ Liq.* ♏x. ; *Aquæ* ℥j. ; thrice daily.

October 18th.—Diet: Meat, green vegetables, potatoes, one slice of toasted bread with each meal. Milk, two pints. Tea or coffee. The dose of fluid extract of cascara was increased to twenty minims.

October 21st.—Ordered, *Ext. Jambolanæ Liq.* (Christy) ʒss. ; *Aquæ* ℥j. ; thrice daily.

This medicine was increased gradually up to two drachms of liquid extract of jambol three times a day, and a euonymin and aloes pill was substituted for the cascara.

October 26th.—She had an attack of diarrhœa, with pain after food.

Ordered, ℞ *Tinctura Opii* ♏xv. ; *Bismuthi Carbonatis* gr. xv. ; *Sodii Bicarbonatis* gr. x. ; *Glycerini* ♏xx. ; *Aq. Chlorof.* ℥j. ; to be given every four hours.

October 27th.—The pulse was intermittent, but the diarrhœa had stopped, though pain was still complained of.

October 29th.—Ordered, *Haust. Magn. Carb.;* thrice daily, and to be made an out-patient.

Her eyes were in the same state on her discharge. There had been no complaint of the pruritus during her stay in the hospital.

The following table shews the state of the urine during her stay in hospital.

DATE.	URINE.	SP. GR.	UREA.	SUGAR.
Nov. 17th.	56	1035	1·8 p. c.	6 p. c.
,, 24th.	46	1040	1·7 p. c.	6 p. c.
,, 26th.	38	1040	—	5·8 p. c.

Jaeger has recorded a case of diabetic *retinitis* where the swelling was so great as to hide the outlines of the disc, and was accompanied by numerous hæmorrhages and yellow patches. There was a marked central *scotoma*. Such central scotomata explain the occurrence of one form of diabetic amblyopia. There is a loss of vision in the central part of the retina, just as in tobacco amblyopia; but cases have been observed in diabetics who were not tobacco smokers (Gowers, Jensen, Nettleship and Edmunds), though they generally are. According to Samelsohn the scotoma may be sometimes peripheral.

Atrophy of the Optic Nerve may be caused by (1) effusion of blood within its sheath, (2) descending degeneration from brain lesions, or (3) ascending degeneration after destruction of the retina by hæmorrhages, (Samelsohn); Nettleship and Edmunds have described a case in which the atrophy appeared to be primary, but the man was a large smoker of tobacco, so that it was not an uncomplicated case.

Capillary aneurysms on the retinal vessels, usually bilateral, have been described.

SMELL AND TASTE.—Jordâo has recorded a case of blunting of taste and smell, but this may have been due to some cerebral complication.

EAR AFFECTIONS.—Deafness is generally due to inflammation of the middle ear, but it may be caused by œdematous swelling of the Eustachian tube (Miot). Diabetic *otitis media* (Griesinger, Jordâo, Külz, Raynaud, Toynbee), has been described as an acute inflammation of the middle ear not dependent upon any external cause. The disease is characterised by intense pain localised in the mastoid region, and is accompanied by tinnitus and intense deafness. There is redness and swelling of the auditory canal with some muco-purulent secretion, and the tympanum is congested, swollen and dull. There is purulent inflammation of the mastoid cells. The disease comes on very suddenly, without any cold or coryza. The osseous tissue is extensively destroyed, and according to Raynaud the disease begins in the bone. Kirchner has described a case of double purulent otitis in a diabetic patient in which there was entire absence of fever.

These cases call for special local treatment on exactly the same principles as in non-diabetic cases, and although the prognosis is not favourable, they may heal very well.

In some cases deafness appears to be purely nervous in its origin, and may supervene at the same time as the commencement of the diabetes.

RESPIRATORY SYSTEM.—The late Warburton Begbie met with a case of *membranous inflammation* of the larynx and trachea in a male patient, aged thirty-nine, under treatment in the Royal Infirmary of Edinburgh for diabetes. The complication proved rapidly fatal, causing suffocation.

The lungs may be affected by *catarrh*, but the most common pulmonary trouble is *chronic pneumonic phthisis* with or without the presence of tubercle bacilli. It comes on very insidiously, often with little cough or rise of temperature, and when discovered there is often a considerable area of lung invaded.

CASE 24.—*Diabetes Mellitus—phthisis—wasting—thirst —polyuria.*

Walter J. B., aged thirty-four, brass worker, was admitted into the General Hospital on July 16th, 1890, complaining of thirst and polyuria. His illness had existed for seven months.

Family History.—Father living and healthy. Mother died, aged thirty-three, of enteric fever. Had two brothers and one sister living, and in good health. There was no history of fits, cancer, diabetes, gout, rheumatism, or insanity in the family.

Previous History.—Patient had diphtheria when he was twenty-three years of age. He had never met with any injury.

Present Illness.—At Christmas, 1889, he found that he was passing an increased quantity of water of pale colour, which left a white deposit when it dried. He was also troubled with thirst, and noticed that he was losing flesh. There was also some irritation about the glans penis. He was under medical treatment for a week. Six weeks ago he began to suffer from cough and pain in his chest, and latterly he spat up some thick yellow phlegm, occasionally tinged with blood. There had been no night sweats. Appetite not increased. Bowels confined. At Christmas, 1889, he weighed 9 st. 10 lbs.

State on Admission.—He was a well-developed but spare man, weighing 7 st. 7 lbs. No jaundice or œdema. Temp. 97·8; Resp. 28; Pulse 108.

Alimentary System.—Lips red and moist; teeth sound; tongue fissured, dry, and coated with yellow fur. Appetite good, but not excessive. No pain or nausea after food. Bowels very constipated. He complained of pain without tenderness across the abdomen below the umbilicus and round to the back. Liver dulness in V. M. L., 4 in.; splenic dulness in M. A. L., 1 in.

Circulatory System.—Apex beat in 5th I. S., internal to V. M. L. Heart not enlarged; sounds normal. Pulse regular and fairly strong.

Respiratory System.—He had a slight cough with scanty expectoration, containing numerous tubercle bacilli. There was flattening above and below both clavicles, and he complained of pain over the manubrium, which was increased by coughing or forced respiration. The percussion note was impaired at the right apex in the supraspinous fossa behind, and as far down as the second rib in front. The breath sounds were harsh, expiration was prolonged, accompanied by moist râles, and vocal resonance and vocal fremitus were increased.

Urinary System.—Urine 126 oz., pale straw-coloured, acid, sp. gr. 1031, urea 1·3 p.c., sugar 6·2 p.c.; no albumen.

In some instances *gangrene* of the lung takes place, from rapid sphacelus of parts affected by the inflammatory process.

CASE 25.—*Diabetes Mellitus—thirst—polyuria—albuminuria—acetonuria—phthisis—gangrene of lung—death.*

James B., twenty-nine, wire drawer, admitted June 27th, 1885, complaining of general weakness, wasting, and thirst.

History.—He had been ill eighteen months, the first symptom noticed being loss of weight. He used to weigh 13st. His work was in an ill-ventilated shop, exposed to acid fumes; he drank about eight quarts of beer daily. His previous health had been very good. His father and mother and five of his brothers and sisters were dead, but he could furnish no precise information as to the causes of death. Three of his sisters were alive and well. An early symptom of his illness was thirst and passing a very large quantity of water; his appetite had latterly been good, but he had got weaker. For a month past he had had a feeling of tightness of the chest and a cough. His eyesight had been unaffected.

Condition on Admission.—He was rather pale, hollow-cheeked and emaciated, weighed 9 stone. His cough was frequent, expectoration about $1\frac{1}{2}$ oz. daily, nummular, greenish grey. Chest flat and hollow below the clavicles. On the right side there was dulness at the right apex with bronchial breathing, metallic crepitations and whispered pectoriloquy; on the right side resonance at apex impaired, breathing harsh with a few crepitations. Pulse 108; Resp. 24; Temp. 101°; urine 298 oz.; pale, clear; sp. gr. 1040; faintly acid; very faint cloud of albumen; 6·89 per cent. sugar; 0·7 per cent. urea; acetone reaction; ferric chloride reaction.

Progress of Case.—He was dieted and treated by extract of opium (gr. i.) thrice daily, and Vichy water. His urine fell to about 100 oz., sp. gr. 1031, sugar 5·8 per cent., but on July 15th, eighteen days after admission, at 9.40 p.m., he was found very weak, with quick pulse, laboured respiration and dilated pupils. He was conscious, and lay quite quietly, but occasionally the right side of his mouth was drawn down by a spasmodic twitching of the platysma myoides. He died at 11 p.m., but he was still conscious shortly before the end.

The *post-mortem* examination was made by Dr. Bull. The body was emaciated; there was no peculiar smell on opening the body, and the blood did not contain excess of fat. The left lung was free, œdematous and congested. No fluid in the pleura. Right lung adherent at upper part, which was converted into a large abscess cavity containing thick sanious fluid, which escaped on removing the lung. The inner surface of this cavity was rugged, dirty green, and a large piece of black gangrenous lung tissue hung by a shred from the posterior wall. Bands of tissue crossed the cavity. Its walls were very soft. It had no very offensive smell. The lower lobe was congested and œdematous, and contained numerous patches of lobular pneumonia, some breaking down. The bronchial glands were greatly enlarged. There was no other special change noted in any other organ, except that the liver and kidneys were very large.

CIRCULATORY SYSTEM.—Valvular disease of the heart, as a consequence of endocarditis occurring in the course of diabetes has been described by Lecorché. According to his statement it usually affects the mitral, rarely the aortic valve. Maguire has met with one or two examples. It is undoubtedly very rare, as in my pathological experience I have only met with this complication once. Affections of the wall of the heart are common, of which the most serious are fatty and fibroid degeneration, which may be attended by attacks of angina pectoris (Vergeley), but which too often attract little attention until death occurs suddenly from syncope.

DIGESTIVE SYSTEM.—Affections of the gums are very common, the most frequent being a form of atrophy in which the teeth loosen and come out; but there is often more or less inflammation present, and the gums may be spongy and bleeding.

CASE 26.—*Diabetes Mellitus—thirst—polyuria—wasting —family history of diabetes—gingivitis—failure of strict diet to remove sugar—some improvement in general condition.*

Thomas C., aged thirty-six, engine-fitter, admitted to the General Hospital on July 16th, 1889, complaining of thirst, polyuria, weakness, and wasting. He had been ill four months, his illness having begun with a slight cold and sore throat, for which he attended the hospital as an out-patient for two months without getting better, when he became very thirsty, his water increased to seven pints daily, he lost his appetite, his tongue was blistered, and his teeth grew loose. He lost weight very rapidly—seven pounds in one week—and had continued to waste ever since. He had been very weak in the legs lately, but the thirst and quantity of water had been less.

Family History.—Father died aged seventy. Mother died aged sixty, of bronchitis. One brother died of diabetes and phthisis. Was married, but wife died eighteen months before of inflammation of lungs. Three children alive and well. No history of gout in family.

Previous History.—Could remember no illness except occasional colds, and a crop of boils twelve years ago. Gonorrhœa eighteen years ago; no history of syphilis.

State on Admission.—Patient was a well-developed sparely nourished man; skin moist; face pale; lips a good colour. Temp. 98°; Pulse 58; Resp. 16. Weight 9 st. 5 lbs.; used to weigh 11 st. 2 lbs.

Alimentary System.—Teeth bad and loose, gums sore; tongue large, white, rather dry. Appetite fair; bowels confined. Liver dulness began in sixth interspace and extended two fingers breadth below the costal border. Splenic dulness normal.

Respiratory System.—No cough or pain in chest, but some dyspnœa on exertion. Percussion and auscultatory signs normal.

Circulatory System.—No palpitation or pain; heart's area not enlarged; apex beat in 5th I. S. inside V.M.L.; sounds normal. Pulse regular, not easily compressible.

Nervous System.—Knee jerks present. Special senses normal. No neuralgic pains or abnormal sensations. Complained of feeling very *irritable* since illness began.

Genito-urinary System.—Total loss of sexual desire. Urine 92 oz.; sp. gr. 1040, acid, pale amber; urea 1·5 per cent., a very faint haze of albumen; sugar 7·4 per cent.; under microscope only bacteria visible.

Treatment.—Diet: Mutton, beef jelly, milk two pints, Vichy water (Haute Rive) and lemon juice. He had no medicine except an occasional aperient. His stay in hospital was only troubled by toothache, which occurred on July 22nd, and lasted some days. He was made an out-patient on September 12th, having been in hospital about two months.

The following table shews the result of treatment.

DATE.	URINE.	SP. GR.	UREA.	SUGAR.	WT. OF PATIENT.
July 17th...	92 oz.	1040	1·5 p.c.	7·4 p.c.	9 st. 5 lbs.
,, 23rd...	96 oz.	1035	2·8 p.c.	6·2 p.c.	8 st. 13 lbs.
,, 30th...	68 oz.	1037	2·8 p.c.	4·6 p.c.	8 st. 12¼ lbs.
Aug. 6th...	90 oz.	1037	—	4·8 p.c.	9 st. 0 lbs.
,, 14th...	100 oz.	1032	2·2 p.c.	3·6 p.c.	9 st. 2 lbs.
,, 19th...	80 oz.	1030	2·4 p.c.	2·6 p.c.	—
,, 22nd...	96 oz.	1036	2·8 p.c.	5·4 p.c.	9 st. 2 lbs.
,, 25th...	106 oz.	1036	—	4·6 p.c.	—
,, 29th...	96 oz.	1036	2·0 p.c.	5·1 p.c.	—
Sept. 1st...	90 oz.	1037	—	5·7 p.c.	—
,, 5th...	96 oz.	1040	1·8 p.c.	5·4 p.c.	—
,, 8th...	98 oz.	1039	2·5 p.c.	5·4 p.c.	—
,, 11th...	106 oz.	1036	2·1 p.c.	4·0 p.c.	9 st. 9¾ lbs.

After becoming an out-patient he continued to gain weight, until he reached 10 stone. He passed about five pints of water daily. In December he complained of profuse sweating. He was treated with various drugs—opium, cocaine, and phosphorus—without any very definite effect; but on the whole he had made progress up to the time he was last seen (April 29th, 1890).

The *mouth* is generally dry, but salivation may be profuse (Rollo). Occasionally complaint is made of hot sour risings into the mouth, causing the teeth to feel as if they were set on edge (Rollo). Pavy has described a sense of emptiness at the pit of the stomach. *Ulceration* of the stomach and intestines may occur, giving rise to vomiting and diarrhœa. The bowels are generally constipated, but *diarrhœa* is sometimes present, generally depending on catarrh.

CASE 27.—*Diabetes Mellitus—rheumatic pains—polyuria—thirst—emaciation—improvement on diet and opium —diarrhœa—failure of jambol.*

Mary P., aged fifty-three, attended as an out-patient on May 4th, 1886, complaining of pains in the back and legs, thirst, loss of flesh, and of passing a great quantity of water. She had been ill four years. The quantity of water in twenty-four hours was about nine pints, sp. gr. 1030, loaded with sugar. She was treated by antidiabetic diet, extract of opium (gr. i.) three times daily, and allowed saccharin as a sweetening agent. On this treatment the quantity of urine fell rapidly to four pints in twenty-four hours, and thirst was lessened. The following year she began to suffer from repeated diarrhœa. Jambol was tried as a substitute for opium, with some temporary rise in the quantity of water. The diarrhœa had to be kept in check by the use of chalk and opium; the urine remained about the same, four pints, sp. gr. 1039, when last seen in September, 1887.

Edwards has related the case of a child, aged seven,

who after suffering from symptoms of diabetes for three or four months, was seized suddenly with pain and tenderness over the abdomen. Examination shewed a considerable quantity of ascitic effusion. The temperature was 102°. His diagnosis was *acute peritonitis* with effusion. Death occurred the same night, but no autopsy appears to have been made.

With reference to the digestion in diabetes, it has been stated by Heller and others that the gastric juice contains sugar. Ponomaroff in a series of experiments in which the œsophageal tube was used and the gastric juice obtained free from bile, found, at no time, that it contained the least trace of sugar. But Rosenstein has found that it often contains no free hydrochloric acid. Where this occurs at intervals he attributes it to a sclerosis affecting the glands, but in other cases it is permanent, and due to atrophy of the mucous membrane secondary to interstitial gastritis. Honigmann disputes this; he found hyperacidity at least as common as anacidity, and refers both to functional causes.

Jaundice may occur as an accidental complication, and on its supervention the sugar as a rule disappears from the urine, but this is not always the case, as shewn in Case 1 (page 15).

Cirrhosis of the liver is more of a pathological than a clinical complication, but enlargement of this organ can be not uncommonly detected on physical examination, and is due to a process of interstitial inflammation which in a certain number of cases leads to atrophy.

ENTERIC FEVER.—There seems to be a more than ordinary liability in diabetics to suffer from enteric fever. Numerous cases have been recorded by Rayer, Griesinger, Bamberger, Gerhardt, Ryba and Plumert, Seifert, Ebstein, etc. The sugar may disappear during the course of the fever. The prognosis is said by Ebstein to be unfavourable.

CASE 28. —*Diabetes Mellitus—polyuria—thirst—wasting—intercurrent pyrexia resembling typhoid.*

Agnes L., eight, school girl, was admitted into hospital on February 7th, 1887, complaining of thirst, hunger, headache, and passing a large quantity of water.

History.—Her mother had noticed that these symptoms had been coming on for six months, as she was constantly drinking water and losing flesh. She had been previously pretty well, but never strong since she was four years old. Her sister was with her in the hospital with diabetes, and another child had died of the same disease at the age of eleven. Six others and the father and mother were alive and well. Nothing was known of any other cases of diabetes in the family.

Condition on Admission.— She was a small, ill-nourished child, of strumous appearance. Temp. 98°, Pulse 96, Resp. 18. Her tongue was dry and clean; the only abnormal signs were deficient resonance and feeble breath sounds at the left apex. Urine 58 oz., sp. gr. 1013, acid, contained 3 per cent. of sugar; no albumen.

Progress of Case.—She was dieted and treated by alkalies and cod liver oil. On this treatment the urine increased in amount, but the sugar diminished and was often absent. She gained weight. But in the middle of February she complained of her throat being sore, and though there was nothing to be seen, her temperature for five days was 101° each evening. During this time the urine fell from 80, 90, or even 130 oz. to 30, 40, or 50 oz. daily, sometimes containing no sugar.

The temperature was normal on the 18th and 19th; on the 20th it rose to 100°; was normal on the 21st; then began to rise, and for the next fifteen days described a course very like that of a typhoid relapse. (*Fig.* 16.) She complained of abdominal pain, her tongue was dry and coated, her face flushed, and the spleen was distinctly enlarged.

On the 28th February, and 1st, 2nd and 3rd March, some very suspicious looking spots were seen on the abdomen, and the stool on one day was said to be quite a typical typhoid stool; but no other spots came out, and on the 8th the temperature was normal. During the period of fever the urine was never more than 44 oz. daily, and sometimes under 20 oz., while there was often no sugar present, and never more than 0·7 per cent.

After this she went on fairly well, but it was difficult to feed her. On her discharge she was improved in appearance, and had gained 6 lbs. in weight.

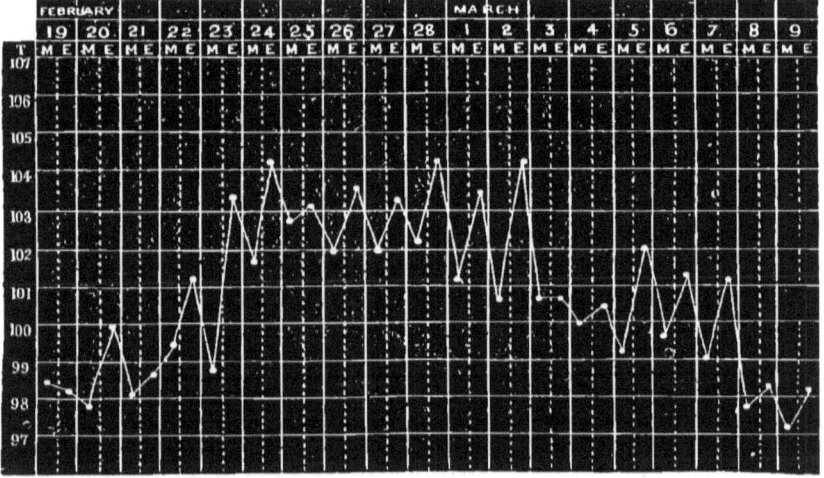

Fig. 10.

RHEUMATISM.—Rheumatic muscular pains are very common in diabetes. In one of my cases taking 1½ oz. of dilute lactic acid daily, profuse sweating and pain occurred, but there was no rise of temperature nor any swelling of the joints.

Thickening and shortening of the palmar aponeurosis (Dupuytren's contraction) has been several times recorded (Cayala, Bordier).

CASE 29.—*Diabetes Mellitus—thirst—polyuria—rheumatic pains—night sweats—acetonuria—enlargement of liver.*

John B., aged fifty-five, wire drawer, was admitted into the General Hospital on September 19th, 1889, complaining of polyuria, wasting, and pain in the morning in the left shoulder. About sixteen months ago he began to complain of pains which he attributed to rheumatism, and his water was greatly increased, the necessity to pass it causing him to rise five or six times in the night. He was very thirsty and drank about a gallon of water or beer daily. A few months later he began to sweat very badly at night.

Previous History.—He had never been very well, and for the last twenty years had suffered from winter cough.

Family History.—Father died aged sixty-eight. Mother died aged sixty, after a paralytic stroke. One brother died of pleurisy; two were living, one of whom suffered from rheumatism and sciatica. A sister died from tumour of the breast, which was removed; two others were living in fair health. He did not know of any case of diabetes or phthisis in the family.

State on Admission.—He was a slenderly-developed, sparely nourished man, with pale cheeks and lips; skin normal, moist; he said it used to itch badly in the early part of his illness, and that three or four of his fingers gathered, but he had had no eruption. Temp. 98°; Pulse, 84; Resp. 14. Weighed 6 stone 12 lbs., and six months before weighed 7 stone.

Alimentary System.—Teeth fairly good, had been loose at times, and occasionally ached; they were much discoloured. Tongue large, covered on dorsum with a moist white paste. Thirst and appetite varied; sometimes they were marked. No pain after food. Bowels regular. Liver dulness in V.M.L. $4\frac{1}{2}$ inches. Splenic dulness in M.A.L. $2\frac{1}{2}$ inches.

Respiratory System.—No cough, pain, or dyspnœa. Percussion note resonant. Breath sounds normal.

Circulatory System.—No pain, palpitation, or dyspnœa. Area of cardiac dulness not altered. Apex beat in 5th I.S. in V.M.L. Sounds normal; pulse regular, full; artery rather hard.

Nervous System.—Eyesight very bad; ophthalmoscopic appearances normal; marked hypermetropia.

Urinary System.—Urine 116 oz., 1035, acid, pale limpid, urea 0·7 per cent.; no albumen; sugar 6 per cent.

Treatment.—Diet: House diet, green vegetables, Vichy water and lemon juice.

Sept. 20th. Ordered *Ext. Opii*, gr. j., three times daily.

Oct. 4th. He was put on a course of *Liq. Arsenicalis*, beginning with 5 minims thrice daily, the dose to be increased 1 minim each day.

Oct. 17th. He was made an out-patient.

The following table shews the progress made.

DATE.	URINE.	SP. GR.	UREA.	SUGAR.	WT. OF PATIENT.
Sept. 19th ...	138	1035	1·1 p. c.	6·4 p. c.	—
,, 22nd ...	84	1035	1·5 p. c.	5·7 p. c.	—
,, 25th ...	139	1031	0·8 p. c.	4·8 p. c.	7 st. 2 lbs.
Oct. 2nd ...	118	1031	2·3 p. c.	3·7 p. c.	—
,, 6th ...	130	1035	2·3 p. c.	5·2 p. c.	7 st. 1 lb.
,, 10th ...	98	1040	1·9 p. c.	4·8 p. c.	—
,, 13th ...	126	1035	1·1 p. c.	4·4 p. c.	—
,, 16th ...	114	1034	1·6 p. c.	4·8 p. c.	7 st. 5 lbs.

During the last fortnight of his stay in hospital his urine gave the reaction of acetone, but this did not have any consequences. He has since been kept under observation as an out-patient, and treated with cocaine, phosphorus, and salicylate of soda, without any decided benefit. On March 25th, 1890, he weighed 6 stone 9 lbs., having recently lost weight, and on careful examination it was found that his liver dulness measured 5½ inches in the V.M.L., passing 3 inches below the costal border.

New Growths.—According to some authorities diabetic patients are specially liable to the formation of tumours, but they are of relatively slow growth (Tuffier).

BIBLIOGRAPHY.

Addison and Gull. On a certain affection of the Skin. Vitiligoidea—a, plana, b, tuberosa, "Guy's Hosp. Reports." 1851.

Althaus (J.). Neuritis of the circumflex nerve in Diabetes. "Lancet," 1890, I, p. 455.

Anderson (McCall). Illustrations of the occurrence and of the gravity of Diabetes Mellitus in early life. "Glas. Med. Jour." Vol. XXVII, p. 344.

Anderson (Wallace). Diabetes Mellitus. "Glas. Med. Jour." Vol. XXV. 1886, p. 126.

Auerbach (L.). Ueber das Verhältniss des Diabetes mellitus zu Affectionen des Nervensystems. "Deut. Arch. für klin. Med." Bd. XLI, Heft. 4, 5.

Babington, quoted by Fagge, *Op. cit.*

Barnes (H.). On exophthalmic goitre and allied neuroses. "Brit. Med. Jour." 1889, I, p. 1225.

Begbie (J. Warburton). A case of fatal croup in the adult. Collected Works, edited by Dyce Duckworth, M.D. Edin. "The New Syd. Soc." 1882.

Bordier (A.). Sur la rétraction diabétique de l'aponévrose palmaire. "Jour. de Thér." 1883.

Bouchard. Pertes des réflexes tendineux dans le diabète sucré. "Gaz. des Hôpitaux," No. 112, 1884, p. 892.

Bristowe (J. S.). Case of Keloid. "Path. Soc. Trans." 1866.

Butel. The relation between Urea, Sugar and Phosphoric Acid in the Urine. "Journ. des Connaissances Méd." 1887.

Buzzard (T.). Illustrations of some less known forms of peripheral neuritis, especially alcoholic monoplegia, and diabetic neuritis. "Brit. Med. Jour." 1890, I, p. 1419.

Cayala (A.). Diabète et rétraction de l'aponevrose palmaire. "Gaz. Hebd." 1883, p. 770.

Cornillon. Des névralgies diabétiques. "Revue de Méd." 1884, p. 213.

Czapek. Beiträge zur Kenntniss der Oxalsaurausscheidung im Menschenharn. "Prag. Ztschr. f. Heilk." Bd. II. p. 345, 1881.

Deutschmann. Pathologisch-anatomische Untersuchungen einiger Augen von Diabetikern. "Arch. f. Ophth." Bd. XXXIII, 1887.

DREYFOUS. On the patellar tendon reflex in Glycosuria. "Revue de Méd." 1887.

DUNCAN (J. MATTHEWS). Translation of Braun on Uræmic Convulsions. Edinburgh, 1857.

EBSTEIN (W.). Weiteres über Diabetes mellitus, insbesondere über die complikation desselben mit typhus abdominalis. "Deutsches Arch. f. klin. Med." Bd. XXVIII, 1881.

EDWARDS (J.). Diabetes Mellitus in a boy seven years of age. "Brit. Med. Jour." 1885, Vol. I., p. 279.

FIENZAL. Paralytic lagophthalmos in Diabetes. "Bull. de la Clin. Nat. Ophth. de l'Hôp. des quinze vingts." Sept. 1885.

FINLAYSON (J.). Diabetes Mellitus: convulsions for seventeen hours: death. "Brit. Med. Jour." 1890. I, p. 1011.

FISCHER (G.) Ueber Beziehungen zwischen Tabes und Diabetes mellitus. "Centrlbl. f. Nervenheilkunde," 1886.

FLINT (AUSTIN). The treatment of Diabetes Mellitus. "New York Med. Jour." 1884, I, p. 588.

FLORAIN. Diabetic Neuralgia. " Gaz. Méd. de Paris " 28th Feb. 1885.

FRANCE (J. F.). Additional notes on diabetic Cataract. "Guy's Hosp. Reports," Third Series, Vol. VI., p. 266. See also "Oph. Hosp. Reports," Jan. 1859.

GALEZOWSKI. Le diabète en pathologie oculaire. "Jour. de Thérap." 1888.

GERARD—., quoted by Rollo, *Op. cit.*

GOODHART (J. F.). Clinical remarks on transient glycosuria of neurotic origin. " Brit. Med. Jour." 1889, II, p. 1381.

GOWERS (W. R.). A manual and atlas of Medical Ophthalmoscopy. 3rd Edition, 1890.

HARKER (J.). Diabetes Mellitus in Children. " Brit. Med. Jour." 1885, I, p. 413.

HARRISON (E.) and SLATER (C.). On the relations of the amounts of Sugar and Urea in the Urine in some cases of Diabetes Mellitus. "Lancet" 1884, I, pp. 338 and 884.

HEUSONER. Ein Fall von Mal perforant du pied. "Deutsch. med. Wochenschr." 1885, No. 16.

HIRSCHMANN (J.). Ein Beitrag zur Lehre von den diabetischen Retinalaffektionen. "Inaug. Diss." Breslau, 1886. "Schmidt's Jahrb." 1886.

HODGKIN (T.). On Diabetes and certain forms of Cachexia, "Association Med. Jour." 1854.

HONIGMANN (G.). Ueber Magenthätigkeit bei Diabetes mellitus. "Deutsche med. Woch." 1890, No. 43.

HUNT (W.). Diabetic gangrene. "Philadelphia Med. News," 1888, p. 687.

JENSEN (E.). Eye Diseases with central Scotoma. *Monograph.* Copenhagen, 1890.

JORDAO. Considérations sur un case de diabète. Paris, 1857.

KAPOSI (M.). Ueber besondere Formen von Hauterkrankung bei Diabetikern; dermatosis diabetica. "Wien. med. Woch." Jan. 1884.

KIRCHNER. Ueber Ohrenkrankheiten bei Diabetes mellitus. Monatsch. für Ohrenheilkunde, 1884, No. 2.

LEBER and WIESINGER. Ueber das Vorkommen von Keratitis, Iritis und Irido-choroiditis bei Diabetes mellitus. "Arch. f. Ophth." Bd. XXXI, 1885, p. 183.

LECORCHÉ. De l'endocardite diabétique. "Arch. Gén. de Méd." 1882.

LE GRAND DU SAULLE. Cerebral complications in Diabetes. "Ann. Med. Psych." 1885.

LE NOBEL (C.). Ueber die Jodformbildender Körper in der Expirationsluft der Diabetiker. "Centrlb. f. d. med. Wis." 1884, No. 24.

LÉPINE (R.). et BLANC. (L.). Hemiplégie diabétique avec lesions seulement microscopiques des circonvolutions motrices. "Revue de Méd." 1886, p. 169.

LEROUX. Diabète sucré chez les Enfants. "Gaz. des Hôp." 1881.

LEUBE (W.). Ueber Glycogen im Harn des Diabetikers. "Virchow's Archiv." Bd. CXIII.

MCBRIDE. PAGE. FOWLER. DRAPER. KINNICUTT. The significance of small quantities of sugar in the urine. Discussion at the New York Academy of Medicine. "New York Med. Jour." XLIII. 1886, p. 353.

MACKENZIE (W.). Practical Treatise on the Diseases of the Eye. 4th Edition, London, 1854.

MADIGAN (H. J.). Insanity and Diabetes. "Journ. of Nerve and Ment. Diseases." Vol. VIII.

MAGUIRE (R.). Albuminuria in Diabetes. "Brit. Med. Jour." 1886, II, p. 543.

MARIE and GUINON. On the loss of the knee-reflex (patellar tendon reflex) in saccharine diabetes. "Revue de Méd." No. 7, 1886, p. 640.

MAX EINHORN. Die Gährungsprobe zum qualitativen Nachweise von Zucker im Harn. "Virchow's Archiv." Bd. CII, p. 263.

MIOT (C.) Réflexions sur l'obstruction de la trompe chez un diabétique. "Revue Mensuelle de Laryngologie," No. 6, 1887.

MÜLLER (F). Pneumaturia. "The Practitioner." Jan. 1890.

MUNK (J.). Zur quantitativen Bestimmung des Zuckers und der sog. reducirenden Substanzen in Harn mittelst Fehling'scher Lösung. "Virchow's Archiv." Bd. CV., p. 63.

NETTLESHIP (E.). Spontaneous disappearance of Diabetic Cataract. "Lancet," 1885, 1, p. 938.

NETTLESHIP (E.). and EDMUNDS (W.). Two cases of symmetrical amblyopia of slow progress with central scotoma, in patients suffering from Diabetes. "Trans. of the Ophth. Soc." Vol. I, p. 124.

OTTO (J.). Das Vorkommen grosser Mengen von Indoxyl und Skatoxylschwefelsäure im Harn des Diabetes mellitus. "Pflüger's Archiv." Bd. XXXIII, p. 607.

PAVY (F. W.). Case of Diabetes, associated with temporary Mania. "Med. Times and Gaz." 1879, II, p. 262.

——————— Introductory Address to the Discussion on the Clinical Aspect of Glycosuria. "Lancet," 1885, II, p. 1035.

PROUT (W.). On the nature and treatment of Stomach and Urinary Diseases. 3rd Edition, London, 1840.

PRYCE (T. DAVIES). A certain cutaneous affection occurring in Diabetes. "Lancet." 1888, II, p. 59.

ROSENBLATH (W.). Ueber multiple Hautnekrosen und Schleimhautulcerationen bei einem Diabetiker. "Virchow's Archiv." Bd. CXIV.

ROSENSTEIN (S.). Ueber das Verhalten des Kniephänomens beim Diabetes mellitus. "Allgem. med. Centralzeitung," 1885, No. 28.

SALOMONSON. Over het ontbreken van den patellair-reflex bij. diabetes mellitus. "Weokbl. van het Nederl. Tijdschr. voor Geneesk." 1890.

SAMELSOHN. Diabetic diseases of the Eye. "Deutsche med. Wochenschr." 1886.

SEEGEN. Ein Fall von Levulose in diabetischen Harn. "Centrlbl. f. d. med. Wissensch." 1884, No. 48.

SEIFERT (O.). Ueber Acetonurie. "Verhandl. der physik.— med. Ges. in Wurzburg." N. F. 4, 1882.

——————— Ein Fall von Diabetes mellitus mit Typhus abdominalis. "Wien. med. Wochenschr." Bd. 31, 1881.

SIMON (J.). Note sur le diabète sucré chez les Enfants. "Revue Médicale," 1885.

SKERRITT (E. MARKHAM). Acute febrile Glycosuria. "Brit. Med. Jour." 1885, II, p. 1052.

SPENCER (W. G.). Diabetes in Surgical Cases. "Westminster Hospital Reports," Vol. IV. 1888, p. 89.

STADELMANN. Ueber die Ursachen der pathologischen Ammoniakausscheidung beim Diabetes mellitus und des Coma diabeticum. "Archiv. f. exp. Path." XVII., p. 419.

STRAHAN (J.). A case of Diabetes with remarks. "Dublin Jour. Med. Sci." Vol. LXXXI, p. 298.

STRUMPELL. Text-book of Medicine, p. 915.

TESCHEMACHER. Ueber wirkliches und scheinbares Aufhören der Zuckerausscheidung bei Diabetes mellitus. "Internationale klinische Rundschau," 1888.

TUFFIER. Diabète et néoplasmes. "Arch. Gén. de Méd." 1888. Vol. II.

VERGELEY (P.). De l'angine de poitrine dans ses rapports avec le Diabète. "Gaz. Hebd." 1883.

WATSON (Sir T.). Lectures on the principles and practice of Physic. 4th Edition, 1857. Vol. II.

WEYL and CITRON. Ueber die Nitrate des Thier und Pflanzekorpers. "Virchow's Archiv." Bd. CI. 1885.

WILLAN.,—quoted by Rollo, *Op. cit.*

WORMS. Le Diabète à évolution lente et son traitement. "Le Progrès Méd." 1889, No. 20, p. 377.

Chapter V.
DIABETIC COMA.

Prout wrote that a diabetic individual may be considered as existing on the brink of a precipice, and since Prout's time the very precarious tenure by which diabetics hold their lives has come to be very generally recognised. We are all aware of the great risks such patients expose themselves to in travelling, probably on account of the unavoidable annoyances, as well as fatigue, which travellers even in these days must put up with. Violent mental emotions and bodily exertion are well known dangers which we cannot be too careful to warn our diabetic patients to avoid.

It is understood that diabetics are more liable than other persons to the same sort of accidents that befal all persons occasionally: thus a diabetic is more likely to have an attack of pleurisy or pneumonia, and such inflammations not unfrequently pass on to breaking down of the lung, or even to gangrene of the inflamed part.

But besides this special predisposition to inflammatory diseases, there is a mode of termination of diabetes which has something characteristic and peculiar in it, and of late years has received a good deal of attention from some of the best clinical observers. It differs from those already alluded to in the alarming nature of the symptoms and the rapidity with which death may supervene in the midst of apparent good health. This it is to which the name of Diabetic Coma has been given, from coma being the final and most constant phenomenon, although it is usually preceded by excitement and dyspnœa, more rarely by nausea and vomiting.

As is well known, diabetic coma was described by Küssmaul, in 1874.

Küssmaul's paper did not meet with the attention it deserved, in this country, until public interest in the subject was aroused by Foster's graphic description of two cases which had come under his observation, in a paper read at the Manchester meeting of the British Medical Association, in 1877. Since that time, many other cases have been recorded, and the general clinical history of the condition has been elaborated by Taylor, Cyr, Lépine, Stephen Mackenzie, Dreschfeld, and others.

ETIOLOGY.—The frequency with which such cases occur may be inferred from the statement of Mackenzie, that of the cases of fatal diabetes collected by him from the registers of the London Hospital, all under the age of twenty-five, with only one exception, had died of coma.

Its relative greater frequency in young persons and acute cases is quite certain, but for the most part its etiology remains obscure. Clinical experience has suggested the dangers of long journeys, muscular exertion, nervous shock, and exposure to cold. Constipation is generally present, is very obstinate, and is theoretically likely to favour the onset of these symptoms.

Bond and Windle in their remarks on a fatal case of diabetic coma which occurred at the General Hospital, say that the change from ordinary diet to richly albuminous food has been present as a factor in this and one other case under their observation. While not doubting the influence which great and sudden changes in diet can produce in the system, and bearing in mind that Charcot states that the converse change frequently produces glycosuria in the novices in the monastery of La Trappe, it is noteworthy that in the case published by Rickards, the patient was taking ordinary diet. This relation to animal diet has not been referred to by other writers, and is obviously absent in many cases which occurred during no treatment at all.

SYMPTOMS.—The premonitory symptoms vary very much. Sometimes the attacks begin with maniacal excitement; more commonly abdominal pain or headache is complained of; a sudden fall in the sp. gr. and sugar contents of the urine has been sometimes noticed, and when observed should be regarded with suspicion, although such an alteration is by no means always a cause for alarm.

Lépine attaches much importance to the rapidity of the pulse as a trustworthy prodromal sign.

The urine gives a Burgundy red colour on the addition of a solution of the perchloride of iron, the colour disappearing on heating the mixture. This reaction is certainly present in diabetic urine apart from coma, and has been met with in other diseases. Thus von Jaksch has met with it in many acute disorders, especially measles, scarlatina, and pneumonia. He has also observed it in a case of gastric cancer terminating fatally by coma. Hoppe-Seyler has described it in a case of sulphuric acid poisoning as occurring during the time no food was taken. Senator observed it in a case of atropine poisoning which died comatose. Windle has observed it in pneumonia, Bright's disease, scarlatina, and several other pathological conditions apart from coma. It cannot be therefore regarded as in any sense a pathognomonic sign, and it was absent in Case 37, page 173.

Acetone is present in the urine, as indicated by a rose-violet coloration with a solution of nitro-prusside of sodium and ammonia. So far as I am aware, this is constant, but the amount present varies greatly. A trace of *albumen* is very commonly to be found.

Another symptom which is very striking when present, but far from constant, is the peculiar odour of the breath, which has been variously described as like sour beer, apples, hay, chloroform, and acetone. Is it probable that all these comparisons have been applied to the same odour? It would seem that acetone or sour beer affects our noses very differently to apples, chloroform, or hay. Unfortunately it is very difficult to institute any precise standard of odour. But Frederick Taylor and Stephen Mackenzie state that they have never been so fortunate as to smell this odour at all. Frerichs groups several of his cases as not presenting this symptom; and it has not been always noticed in the cases we have observed at the General Hospital.

These differences indicate the importance of observing and recording carefully all the facts in these cases when they come under observation.

It is probable that there may be two or more types of sudden death in diabetes, and that these may be dependent upon an equal number of distinct morbid influences; but in order to determine this we require first of all carefully recorded clinical histories.

For want of clinical data all pathological work comes sooner or later to a standstill. Morbid anatomy with all its refinements and aids from histology and chemistry is not, and can never be, a substantive science. It is the great adjunct to clinical medicine, but the latter must lead the way.

Frerichs attempted to give greater precision to our study, by classifying cases of sudden death in diabetes into three groups.

In the first group he placed cases which suddenly, and usually after previous exertion, become prostrated, with cold extremities, small failing pulse, drowsiness, and loss of consciousness, terminating fatally in a few hours.

In the second group the duration is longer, and there is a prodromal stage which may be general prostration, gastric disturbance, vertigo, vomiting, constipation, or a local disease, a dental abscess, pharyngitis, an abscess with a tendency to gangrene, bronchitis, or catarrhal pneumonia. The attack itself commences with headache, restlessness, delirium, anxiety, sometimes with maniacal outbursts, dyspnœa, frequent deep respiration with free entrance of air into the lungs, sometimes with cyanosis, sometimes without it, feeble rapid pulse, low temperature, drowsiness and coma. The breath has a peculiar smell like fruit or chloroform, or acetone; Frerichs is not very definite about the smell. Such cases may recover temporarily, and in some rare instances the attack may quite pass away. The duration of the symptoms may be from twenty-four hours to three or four days, or even longer.

In the third group he placed cases which present no dyspnœa or anxiety, have moderately firm pulses, and are fairly well nourished. The attack is characterised by headache, a feeling of intoxication, with disordered gait, sleepiness, and gradual coma, from which they do not awaken. The breath has the characteristic smell.

In the first group the symptoms are those of collapse, coma occurring only at the end, and the duration of the whole attack is very short. In the second group we have Küssmaul's typical complex of symptoms, with dyspnœa, peculiar odour of breath, and coma. In the third group there is no dyspnœa, the symptoms more closely resemble intoxication, but the breath has the characteristic smell, and coma is present.

PATHOLOGY.—Our knowledge of the pathology of this subject has been much widened, and our view of it has gained much in comprehensiveness from the writings of Senator, who has sought to establish the existence of a self-infective process, dependent upon the formation of toxic substances in normal or pathological cavities of the body, which, occurring in many other conditions besides diabetes, gives rise to phenomena essentially identical with those described by Küssmaul, and corresponds to Frerichs' second group. He has recorded seven such cases; two of chronic cystitis, two of gastric cancer, and three of pernicious anæmia. In none did the urine contain any sugar, or *give a reaction with ferric chloride*. Riess has described eight cases in anæmia, five in anæmia with renal disease, and four in gastric and hepatic cancer. Von Jaksch has published a case of " coma carcinomatosum," in a patient the subject of gastric cancer, the urine containing acetone and aceto-acetic acid but no sugar. Litten has described one case which occurred under what he calls " dyspeptic conditions," and which terminated in recovery, but the patient was a boy convalescing from scarlatina, and suffering at the time from albuminuria.

I am, therefore, disposed to accept the view that Küssmaul's coma, if I may be permitted to use a convenient though inexact term, is not restricted in its occurrence to cases of diabetes, but may be met with in several other diseases, especially in those in which the state of the blood has undergone profound pathological alteration.

This is proved by the case of Harriet B., which is published at length on page 69 of my " Lectures on Bright's Disease." In her case, the cause of the coma was pyonephrosis with renal calculus, and there was no glycosuria, but the type of the coma corresponded to Küssmaul's description.

Various theories have been brought forward to account for these symptoms. As is well known, Küssmaul adopted the view that the phenomena depended upon poisoning of the nerve centres by acetone.

Acetonæmia was not altogether a new idea in pathology, as Petters, as long ago as 1857, had found acetone in the blood, expired air, and urine, of a severe case of diabetes. Later on Kaulich noticed an acetone-like smell in the urine of patients suffering from variola, typhus, and pneumonia. The presence of acetone in diabetic urine was subsequently confirmed by Cantani, Kaulich, Rupstein, and Fleischer.

The Ferric Chloride Reaction.—Gerhardt observed in a case of diabetic acetonuria that the urine gave a peculiar reddish-brown colour with ferric chloride, which he regarded as characteristic of the presence of acetone. He ascribed this reaction to the presence of diacetic ether, which easily broke up into acetone, alcohol, and carbonic acid. Rupstein succeeded in obtaining the ferric chloride reaction with the ethereal extract of such urine. But the reaction sometimes failed in diabetic urine which undoubtedly contained acetone. Salkowski obtained the reaction eight times in fourteen cases of diabetes. The colour disappeared on heating and acidulation, but with no smell of acetone, nor was there any smell of acetone about the urine or the patients themselves. He never succeeded in getting the ferric chloride reaction with the ethereal extract.

Fleischer found that fermentation with yeast did not destroy the power of giving this reaction in diabetic urine, but did destroy it in urine to which diacetic ether was added, so that he concluded, and his conclusion has been generally accepted, that this latter substance is not that which gives the ferric chloride reaction.

Von Jaksch then suggested that the reaction was due to aceto-acetic acid, which readily breaks up into acetone and carbonic acid; but this view has been recently forcibly contested by Le Nobel, who points out that while this acid is so unstable that it cannot be kept even a few hours in a stoppered bottle without undergoing decomposition, yet the substance which gives the ferric chloride reaction may be extracted from the urine by ether, and in no instance has he succeeded in obtaining from it chemical proof of the presence of acetone, which, if it were in truth aceto-acetic acid, should rapidly be formed.

This particular part of the enquiry therefore stands thus:—When a solution of ferric chloride is added to urine, in certain cases a reddish-brown or Burgundy red colour is produced, which disappears on heating or acidulation, but what this substance is must be considered as still an open question; it is neither acetone nor diacetic ether, and it is not certain that it is aceto-acetic acid.

In making use of the ferric chloride test it is necessary to remember that heat darkens a solution of ferric salts, so that though heat may dispel the colour caused by the presence of the abnormal urinary constituent, the fluid will be finally darker than it would be by the addition of the ferric chloride to normal urine, without heating.

It is very important not to omit the application of heat, as the following substances, any of which may be present in the urine, give the same colour reaction, distinguished only by its persistence on heating. These are:—β-oxybutyric acid, sulpho-cyanides, formic acid compounds, acetic acid compounds; while carbolic and salicylic acids give a blackish brown coloration, not always distinguishable except by the heat test.

Besides these chemical difficulties, clinical observation has shewn that this reaction has a very wide-spread range of occurrence, and that its presence *per se* is of no special clinical significance. I have briefly summarised some of these conditions :—

1.—It has been observed in many acute diseases, *e.g.*, measles, scarlatina, and pneumonia, without any symptoms of coma being present.

2.—It has been noticed in a variety of other conditions, such as cancer, chronic Bright's disease, perityphlitis, strangulated hernia, after minor surgical operations, and in sulphuric acid poisoning, without any dyspnœa or coma.

3.—It has been present in the urine of diabetics for weeks and months without coma or dyspnœa supervening.

4.—It has been observed in the urine of a patient dying from coma with cancer of the stomach, in whose urine acetone and diacetic acid were present, but no sugar.

5.—It was observed by Litten in a post-scarlatinal patient, who presented symptoms resembling those described by Küssmaul. Litten speaks of the condition as "dyspeptic," but albuminuria was present, and the case was complex.

6.—It has frequently been observed in the urine of diabetics during or just before the onset of the peculiar terminal dyspnœa.

Acetonuria.—There is very little doubt that acetone is frequently present in the urine of diabetics, for though chemical investigations of this sort are most difficult, I think we must accept this as an undoubted fact. The best test for acetone has been discovered by Le Nobel. This depends upon the peculiar colour reaction produced by the addition of a solution of nitro-prusside of sodium and ammonia to the urine, or any other fluid containing acetone. A characteristic rose-violet colour is slowly developed, but its appearance is hastened by acidulation, or by shaking the mixture with air.

The older test depended upon the formation of iodoform in the presence of acetone, when iodine and iodide of potassium and ammonia are added to the urine or other fluid. There is another which depends on the solubility of mercuric oxide in the presence of acetone, a salt which is very insoluble in water or alcohol.

Acetonæmia.—The blood of Sarah L. (Case 32, see page 158), was distilled very carefully, but no acetone could be detected in the distillate by the same tests. These investigations were conducted in Prof. Tilden's laboratory, where I had the great advantage of his kind assistance and co-operation.

A great difficulty in the way of accepting the view that acetonæmia is the cause of these toxic symptoms, is that it has not been proved that acetone is capable of giving rise to similar physiological effects. Küssmaul obtained results which were of a not very decided character, but more recently Salomon and Brieger have shewn that acetone in large doses produces no effect on animals or men, even on diabetic patients; while the urine of the subjects of these experiments had no smell of acetone, gave no reaction with ferric chloride, nor any of the chemical reactions of acetone, so that it appears that acetone is destroyed in the body.

Similar experiments by the same observers with aceto-acetic acid, caused acetone to appear in the urine, where it announced its presence by its smell, the iodoform reaction, and its characteristic combination with bisulphide of soda, but in no case did the urine give the ferric chloride reaction.

Very large doses were without any noticeable effect upon the general condition of the subjects of the experiment; neither dyspnœa nor somnolence was produced; there was solely some loss of appetite after the doses had been continued for many days, and the breath had a peculiar aromatic smell.

Penzoldt has stated that when excretion through the lungs is retarded, the introduction of large quantities of acetone into the circulation of rabbits, is followed by intoxication, hebitude, and coma, and he contends that when the lungs are disabled from any cause, these results will follow acetonæmia in man. Unfortunately for the application of these facts to the explanation of the cases under consideration, the absence of pulmonary complications has been specially noted in the majority of them, and has even formed the basis for an aphorism, viz., " that when pulmonary disease is absent or slight, the occurrence of coma is more to be feared, especially the younger the patient and the more acute the disease " (Mackenzie).

Moreover, it is known that acetonuria may occur independently of diabetes and without being followed by any comatose symptoms. Thus, Bull, who investigated this subject at my request, found it present in five out of twenty-one cases of pneumonia, in acute chorea, tonsillitis, intestinal colic, and various surgical conditions, such as fractures, burns, and contusions. After the ingestion of alcohol acetone may be found in the urine, but this fallacy was known and did not account for the acetonuria in the above cases.

Senator in the paper already alluded to suggests that *tri-methylamine* may be the toxic agent, but advances little evidence in favour of this hypothesis.

Minkowski has discovered the presence in the blood of diabetics of large quantities of an acid, which he believes he has identified as *β-oxybutyric acid*, one of the isomeric series of butyric acids, and has introduced a new theory by calling attention to the toxic influence of large quantities of acids when introduced into the body. Walter, in the course of some researches into the effects of *acids* on the animal organism, noticed that in rabbits the introduction into the stomach of large quantities of diluted phosphoric and hydrochloric acids was followed by "dyspnœa, depression of the heart's action, and death by collapse." *Post mortem* examination shewed in some cases erosion of the mucous membrane of the stomach, and the coagulation of the blood was delayed. The objection that the symptoms might have been due to the local action of the acids on the stomach, was disproved by the effects of *subcutaneous* injections of alkalies, which prevented or cut them short. From his investigation, he arrived at the conclusion that the de-alkalisation of the blood by the introduction into the body of excess of acids, causes first stimulation, and later on paralysis of the respiratory centre.

It is important to notice that this β-oxybutyric acid is capable of breaking up to form aceto-acetic acid, which as I have previously stated, yields on decomposition acetone and carbonic acid, so that this substance must be regarded as being nearly allied to acetone.

Binz has found that sodium butyrate produces coma in cats.

Latham has suggested the following theoretical explanation of the production of a toxic substance in the blood He thinks that the second cyan-alcohol $C_2H_4 \begin{cases} OH \\ CN \end{cases}$ is converted by hydration into lactic acid, which if not completely oxidised into CO_2 and H_2O would first form aldehyde, thus:—

$$C_2H_4 \begin{cases} OH \\ COOH \end{cases} + O = \begin{cases} CH_3 \\ COH \end{cases} + CO_2 + H_2O.$$
Lactic Acid.   Acetic Aldehyde.

Then by condensation:—
$$3\,C_2H_4O = C_6H_{12}O_3.$$
Paraldehyde.

This is a hypnotic, and would cause drowsiness.

Or he suggests that the third cyan-alcohol $C_3H_6 \begin{cases} OH \\ CN \end{cases}$ might be hydrated into oxybutyric acid.

Lipæmia and Fat Embolism.—In spite of the negative results of most *post mortem* examinations, an attempt was made by Sanders and Hamilton to find a structural basis for these symptoms. Their view, as is well known, is that the respiratory and nervous phenomena are caused by fat embolism of the pulmonary and cerebral capillaries. They based this theory on the results of the *post mortem* examination of one case of diabetic coma, in which the blood was very fatty, and fat embola were found in the lungs and kidneys.

It is curious that Küssmaul's first case had also fatty blood, and the lungs presented numerous small infarcts respecting which he remarks, "perhaps we might attribute to the lipæmia and fat embolism the numerous small lung infarcts, at least another source of embolism was not to be discovered. *In no case could the fearful terminal dyspnœa have originated in these small infarcts, the greater part of which were of far older date.*" Thus this theory had not escaped the notice of Küssmaul, although he merely noticed it to dismiss it. Lipæmia is only exceptionally present in diabetic coma, and even when it exists, fat embola have not been found (Taylor, Dreschfeld), while in two cases with lipæmia, examined by Mr. Barling and myself, the fat in the vessels had assumed rather the appearance of *post mortem* thrombi than of embolism, and these were fewer than we saw in many cases of fracture where they had given rise to no symptoms at all during life.

Uræmia.—Although Stockvis and even Ebstein, maintain the opinion that these symptoms are of a uræmic nature they have found few supporters; clinically, they differ from the classical type of uræmia, in which convulsions play the leading rôle; etiologically, neither suppression of urine nor diminution in the normal urinary solids is a constant phenomenon; finally, the doctrine of uræmia rests on quite as uncertain a basis as that of acetonæmia.

In concluding this part of the subject, I may express my own opinion, that, while at present we are unable to determine positively the nature of the toxic substance, or the determining causes of the sudden explosion of the fatal terminal symptoms, these phenomena are toxæmic, and in diabetes depend upon the presence in the blood of some substance nearly allied to acetone.

Heart Failure.—There has never been any dispute that *heart failure* is the obvious explanation of a certain number of these cases, such, for example, as Frerichs places in his first group. This is said by Schmitz to be due to *fatty degeneration* of the heart, and there can be no doubt that this is the usual form of cardiac degeneration in diabetes. Frerichs has suggested that the muscular fibres of the heart in diabetes undergo a "glycogenic degeneration," in which their power of contractility is lost, yet it is doubtful if the deposit of glycogen really injures the heart's muscle.

The following case is an example of this type, where the cardiac muscle had undergone brown atrophy:—

CASE 30.—*Diabetic coma—death from heart failure.*

F. R. W., an engine driver, aged forty-one, unmarried, was admitted on May 17th, 1884; he had been attending as an out-patient since the previous December.

His own account of himself was that he had always been a healthy man; he came of a good stock, his mother and several brothers and sisters being alive and well, and his father having died of smallpox at the age of fifty-four. About a month before he was first seen, he was leaning over the strap of the fly-wheel cleaning his engine, when his foot slipped, and he fell so as to contuse his abdomen and strain his back. He came at once to the hospital, and was treated surgically for the strain, but got weaker, and was eventually transferred to my care.

At that time he had lost a good deal of flesh, he complained much of weakness, and passed a large quantity of water, which was loaded with sugar, but contained no albumen, and gave no special colour reaction with ferric chloride. On modified diet and treatment he improved somewhat: that is to say, the quantity and sp. gr. of the urine were reduced; but he continued to lose weight, and became so very anæmic and weak, in spite of iron and cod liver oil, that he was admitted into the hospital. He had slight signs of phthisis at the left apex. After admission he improved a little, gaining 5¼ lbs. in weight by June 14th. His bowels were very obstinately constipated, and towards the end of June he had some gastric trouble for which all his medicines were stopped except an occasional purgative dose of jalap.

On July 4th, he complained of faintness; as we were on the look out for the supervention of coma, his urine was tested with ferric chloride and found to give a marked colour reaction which disappeared on heating.

On July 8th he went out to see his mother who lived in the town; the next day he complained of feeling very tired and did not eat his food. On that day also the ferric chloride reaction was present.

On the evening of July 10th he said he was feeling very tired, and complained of sharp pain in the left side of the chest with difficulty of breathing. This was relieved by hot fomentation. The following day he was still tired and *drowsy;* his hands and feet were cold. Towards evening *his respirations became deep, and sighing,* he was cold and drowsy, but could be roused, and answered questions intelligibly. He died quietly at 7 a.m.

The *post mortem* examination was performed the same day by Dr. Foxwell. The brain weighed fifty-nine ounces; there was a cyst the size of a horse bean in the white matter of each frontal lobe, and the lateral ventricles were dilated and contained clear fluid. At the apex of the left lung were two small cavities, and the lower lobe was in a state of recent pneumonic consolidation. There was no unusual smell on opening the body, but the blood which ran out as the organs were removed was of a milky purple colour. On standing it looked like black currant juice on which cream was floating. Under the microscope the cream-like layer consisted of finely molecular matter, entirely soluble in ether. Some of the blood was distilled by Dr. MacMunn, but no *acetone* could be detected in the distillate. The liver weighed 81oz., it was soft, friable, and its cut surface was mottled with yellow patches. The spleen weighed 5 oz., and appeared normal. The pancreas weighed 1½ oz., it was only 4 inches long, and very friable. The kidneys weighed together 18 oz., they were coarse, pale, and soft, their capsules stripped off readily. The supra-renal capsules were healthy. The semi-lunar ganglia were much enlarged, and of dense consistence. The large intestine was stuffed full of solid fæces.

The semi-lunar ganglia were hardened in picric acid and examined microscopically. The sections shewed great increase of connective tissue which was swollen and hyaline, and crowded with lymphoid cells. The nerve fibres were abundant, and their nuclei well marked. The ganglionic cells were loaded with pigment but otherwise unchanged.

Two or three years ago I made a careful study of the semi-lunar ganglia, and I am therefore able to speak with confidence as to the nature of these changes in organs whose histology is not very familiar. There is no doubt that alterations in these and other sympathetic ganglia are often secondary to diseases of the viscera with which they are functionally related. But although the liver and kidneys presented some fatty change, I cannot regard that as sufficient to account for the lesions of the ganglia. Certainly in no case of Bright's disease, whether acute or chronic, have I seen ganglia so large as these, and I am strongly inclined to look upon the immense overgrowth of ganglionic connective tissue as a primary structural disorder, and to connect it with the blow on the abdomen which seems to have been the starting point of the case.

The lungs, liver, and kidneys were examined microscopically for fat embola. The lungs presented no branching embola, but in some cases the capillaries contained droplets of fat large enough to occlude the lumina of the vessels. The liver capillaries contained numerous similar droplets. The connective tissue of the liver around the ducts shewed some infiltration with round cells. The kidneys were very slightly fatty, some tubules contained hyaline casts, and there was some increase of connective tissue around the larger veins. The capillaries contained very few fat droplets on the whole; by far the largest amount of fat was seen in the capillaries of the liver, and this to such an extent as to suggest very strongly that it was formed there.

I examined the heart very carefully, following Frerichs' directions, but the muscular fibre, though shewing a considerable degree of pigmentary or brown atrophy was free from any evidence of glycogenic change.

In the second group, Küssmaul's typical complex of symptoms is present—dyspnœa, peculiar odour of breath, and coma. It is this group which corresponds to Küssmaul's description. He wrote: " At my first visit to the patient, at 11 o'clock in the morning, I found her in bed, but very restless, tossing about in mortal fright, and imploring assistance. She looked very pale, her face and body were cool, the extremities cold, pulse very small, easily compressed, very rapid (135-140). Respiration quick (36), and respiratory movements very great. Violent costo-abdominal inspiration alternated with violent expiration; expansion of the thorax took place in every direction. She complained of great constriction with very severe pain in both sides of the hypogastrium, and she remarked that it was about her menstrual period. The abdomen was soft and everywhere could be deeply explored without anything unusual being perceived. Deep pressure in the hypogastrium was painful. The heart's sounds were feeble; the breath sounds loud, pure, unaccompanied by any sibilus, crepitus, or râle. The intellect appeared quite clear. She complained of great thirst and drank much spring and Vichy water. She passed a quantity of urine of a straw colour, which contained a large quantity of sugar but no albumen."

In his summary of symptoms, Küssmaul speaks of the following: (1) a peculiar dyspnœa; (2) rapid heart's action;

(3) groaning, screaming, jactitation, and pain; (4) normal or subnormal temperature; (5) abundant urine containing sugar; (6) contraction of pupils (in 2 cases); and (7) paralytic distension of the stomach. Curiously there is not a word about coma, or of any peculiar smell of the breath.

Frerichs' second type is illustrated by the following case:—

CASE 31.—*Acute Diabetes—coma—death—autopsy.*

F. W. A., seventeen, jeweller, presented himself as an out-patient at the General Hospital, on Thursday, Oct. 19th, 1882, complaining that he was rapidly losing strength and flesh, and passing a very large quantity of urine. He looked very ill, and his urine, on examination, contained a large quantity of sugar, was of sp. gr. 1040 but gave no reaction with ferric chloride. He was recommended for admission, and the following history was obtained. There was no known instance of diabetes in the family; his mother died of phthisis three months ago, and this circumstance had been a source of great grief to him. His previous health had been good, with the exception of bilious attacks, to which he was subject. Five weeks ago, he began to feel ill and to pass water during the night, and for the last fortnight he had been passing very large quantities both in the day and night, altogether, he estimated, about two bucketsful of what he described as "hay-smelling" urine. For the last two weeks he had suffered from constant frontal headache, and great thirst; his appetite had continued good, but not inordinately so. His bowels during the same period had been open every other day. Five days ago he vomited, and had to give up his work. His penis became a little sore, and he noticed that his urine left a sticky stain on his shirt. He had been taking medicine from a chemist for four days before admission.

Present Condition, October 20th, 11 a.m. (the day after admission):—Patient complained of frontal headache, weakness, constant desire to pass water, and great thirst. He was an emaciated looking lad, with dilated pupils, and dark rings under his eyes; his alæ nasi were working, but his lips were of a good red colour. Expression anxious; skin dry and rough; Temp. 98°; Pulse, 112; Respirations, 24. No œdema of legs. Teeth glazed; gums red and slightly spongy; throat covered with thick mucus; fauces reddened; tongue very dry, and its centre coated with a grey fur; appetite fair, no vomiting. Bowels were opened on the day before admission (October 18th). His breath smelt sweet; there was no cough; breathing deeper than in health. The circulatory organs appeared normal. During the night he passed a large quantity of urine, which was acid, sp. gr. 1040, contained $\frac{1}{8}$ column of albumen, and a large quantity of sugar. The urine passed in the morning was acid, clear, greenish coloured, sp. gr. 1035, contained $\frac{1}{8}$ column of albumen, a large quantity of sugar, and gave an unmistakable deep vinous-coloured reaction with ferric chloride. His eyesight was unaffected. On the day of admission the patient took ordinary diet, but since he was on diabetic diet.

The detection of the ferric chloride reaction at once suggested the fear that coma might be imminent, and as the bowels had not been moved for forty-eight hours, two scruples of pulv. jalapæ co. were administered at 12.30 p.m., and he was ordered to be carefully watched.

3 p.m.—Patient became very restless, and complained of great pain in his abdomen; his respirations were deeper, and his pulse quick and feeble. The pain in the abdomen was relieved somewhat by hot fomentations. The bowels had not been moved.

5 p.m.—Pulse 156, feeble; Resp. 34. Ordered a teaspoonful of brandy every half hour.

5.30 p.m.—No cyanosis; respirations deep and sighing; pulse very feeble; complained of no pain. Not restless, seemed rather drowsy. Ordered three teaspoonfuls of brandy at once, and a teaspoonful every ten minutes.

7 p.m.—Patient quite comatose; thirty minims of ether injected subcutaneously did not rouse him, nor did he appear to feel the operation of having a drop of blood taken from his finger. After a little he roused sufficiently to put out his tongue and to say that he felt short of breath, but no pain. There was no cyanosis; breathing was very deep, 41; pulse (before the ether) 174, extremely feeble. Pupils were dilated, but reacted to light. Breathing very harsh; heart sounds scarcely perceptible. The blood on examination did not contain any fat globules, but the leucocytes appeared slightly increased. An enema of one ounce of brandy and a drachm of ether, in two ounces of water, was then administered, and he was enveloped in a hot pack.

9 p.m.—Patient had been very violent, tossing himself about in the bed; radial pulse not to be felt; pupils dilated and insensible to light. Breathing deep; no cyanosis. He was deeply comatose and could not be roused. Later on his father managed to rouse him by repeated shouting, and on his father saying, "Do you hear me? patient replied, "Yes! I hear you."

11 p.m.—Patient died; eight hours after the onset of the more serious symptoms. The nurse reported that he turned "bluish" and struggled very violently before his death.

No urine was passed after 6 p.m.; but ten ounces were passed during the previous hour; this was examined by Dr. Windle, and he reported it to have been pale yellow, clear, acid, sp. gr. 1030, containing a trace of albumen, 1·6 per cent. of sugar, ·95 per cent. of urea; chlorides diminished; no trace of indoxyl; a deep vinous red reaction with ferric chloride.

The total quantity of urine passed in the twenty-four hours, from 5 p.m., October 19th, to 5 p.m., October 20th, was 160 ounces. The temperature on the evening of the 19th was 100° Fahrenheit, but fell gradually, being 98° on the morning of the 20th, and 97° at 7 p.m., four hours before his death.

AUTOPSY.—The *post mortem* examination was performed by Dr. Windle, on October 21st, fifteen hours after death.

External Appearances.—The body was that of a young male, somewhat emaciated; cadaveric rigidity and hypostatic congestion were well marked.

The *Spinal Cord* presented no abnormal naked eye appearance.

Head.—The cranial bones were very thin; the dura mater was extremely adherent; there was no clot in the superior longitudinal sinus. There was increase of the Pacchionian bodies on the arachnoid along the longitudinal fissure. The brain substance was firm, slightly hyperæmic throughout, with numerous puncta cruenta; a small amount of reddish serum in the lateral ventricles.

Thorax.—The cavities of the heart contained some dark fluid blood, and a very few soft clots; the valves were normal; the heart substance was rather pale. The lungs, when squeezed, emitted a peculiar sour acetone-like odour, which was also observed on section of them. Their substance was slightly hyperæmic, but otherwise normal.

Abdomen.—The stomach and intestines were much distended. There was no unusual smell detected on opening this cavity. The diaphragmatic concavity reached as high as the upper edge of the fifth rib. The liver was slightly enlarged, and on section pale in colour. The gall bladder was almost empty and collapsed. The bile ducts were pervious. The spleen was small and soft. The kidneys were rather large, their capsules stripped readily; substance pale. The supra-renal capsules were normal. The pancreas was small and shrunken in appearance. The stomach was full of pultaceous food; its mucous surface shewed patches of hyperæmia. The intestines were loaded with fæces, the sigmoid flexure and rectum being stuffed full of hard grey masses, quite uncoloured by bile. The small intestine was full of pale yellow soft fæces. There was a large mass of tænia in the colon. The solar plexus and cervical sympathetic ganglia appeared normal. The blood nowhere presented any abnormal appearance.

On microscopical examination the liver, spleen, kidneys, and lungs were quite normal. There were no fat embola or any traces of fat in the vessels of the lungs and kidneys, or in the renal tubules. The medulla was also normal, though the perivascular spaces were slightly enlarged. The cells of the pancreas were badly defined, and their contents cloudy. The sympathetic ganglia were unfortunately lost.

Dr. Norris, who examined the blood, reported that the leucocytes were peculiarly large, and the red corpuscles and granular matter slightly in excess, but there was no excess of fat.

The clinical phenomena presented by this case are of great interest. In the first place, the diabetes was apparently produced by nervous shock, the profound grief induced by the death of a near relative, a fact which is in accordance with many previous observations, and which goes with them to support the view of the purely nervous origin of the disease. Secondly, the course of the disorder was remarkably acute and rapid. Thirdly, we were able to trace the development of the whole of the terminal symptoms, and to foresee their approach by the use of the ferric chloride test.

In considering the predisposing causes of this special termination of diabetes, we must place first the *age* of the patient, and the *acuteness* of the disease. The important fact that it is in *young persons* and in *acute cases*, that diabetic coma is mostly liable to supervene is now well established, and is supported by the statistics of Guy's Hospital, adduced by Taylor.

In more immediate and direct relation to the actual attack was the *constipation*, which the *post mortem* examination proved to have been much greater than I had supposed from the statement of the patient that his bowels had been opened every other day. I was fully aware of the sinister influence of this condition, and it was in the hope of removing it, and averting the danger which I apprehended, that the compound jalap powder was ordered, but unfortunately without effect.

Besides constipation, nervous shock, sudden or prolonged muscular exertion, cold, and the fatigue of travelling, have been recorded as predisposing factors, but none of these were present in this case. The patient lived in the town, he had only a short distance to come to the hospital, and there is no reason to believe that that circumstance in itself disturbed him at all.

Among the earliest of the premonitory signs was the alteration of the respiration: in the morning note we find that the breathing was "deeper than in health."

Throughout the report of the clinical symptoms we have repeatedly the note, "no cyanosis," and until just before death, when according to the nurse he turned "bluish," there is not the slightest evidence of any alteration in the function of blood aëration. Undoubtedly cyanosis has been described in other cases, but as it is not constant, when present it is probably due to some complication. Its absence indicates that the dyspnœa is purely nervous in origin, and the result of stimulation of the respiratory centre, rather than of any local change in the lungs, and affords an argument against the theory of Sanders and Hamilton that the dyspnœa is due to interference with the circulation through the lungs, by fat embolism of the pulmonary capillaries.

The *pulse* at the morning visit was 112, and soon rose to be extremely rapid and feeble; this symptom is regarded by Lépine as being of great value as an early indication of the onset.

Abdominal pain was complained of first at 3 p.m., and may have been partly the effect of the purgative. It is, however, known to be a very constant and early sign in these cases.

Drowsiness was first noticed at 5.30 p.m., and by 7 o'clock he was quite comatose.

Death was preceded by *convulsions*. Senator speaks of twitchings as sometimes supervening, but in another case under my care as in this, actual convulsions occurred. Convulsions have also been observed by Minot and by Buhl. Cyr considers that the absence of convulsions affords a diagnostic distinction between this form of coma and that from uræmia, but this, it is obvious, cannot be maintained.

Among the premonitory signs presented by this case, and which I regard as of great value, were the gradual fall in the sp. gr. and quantity of the urine, and consequent diminution of the amount of sugar excreted, which was far too great to be accounted for by the change of diet. In the last sample of urine examined the quantity of sugar was only about seven grains to the ounce, an amount which must have been greatly surpassed in the previous day, when the sp. gr. was ten degrees higher. A sample of the urine, some days old, was sent to Prof. Tilden, F.R.S., at Mason's College, to be distilled for acetone, but he reported that none was to be found.

The percentage of urea in the last urine was low, being under one per cent., but considering the large quantity of urine passed, the total amount excreted during the final twenty-four hours of life must have been quite seven hundred grains. This disposes of the objection that the toxic phenomena were due to non-elimination of urea. No urine was passed after 6 p.m., but this coincided with the supervention of the coma. Ebstein's suggestion that necrosis of the renal epithelium causes the retention of sugar and urea in the blood is, moreover, opposed in this case, by the negative results of the microscopical examination of the kidneys.

This case illustrates very strikingly the prognostic use that may be made of the ferric chloride reaction, for although it is not uncommonly found in diabetes without being followed by coma, and is often present in the urine of other diseases, yet its sudden appearance in the urine, especially in young subjects and in acute cases of diabetes, should be regarded as a sign of grave moment, and should direct attention to the state of the bowels, the respiration, the pulse, and the other prodromata which have been described.

In the third group, there is no dyspnœa, the symptoms more closely resemble intoxication, but the breath has the characteristic smell, and coma is present. This third group is relatively rare; it is characterised by the *absence* of the peculiar dyspnœa. I have never met with an example of it. Frerichs records three cases, and I think the case of diabetic coma reported by Prescott Roberts was of this kind; but his account is very brief.

In all these forms the *temperature* as a rule is normal, or subnormal, but it may rise to a considerable height.

CASE 32.—*Diabetes Mellitus—death from coma with high temperature.*

Sarah L., nineteen, tailoress, was admitted on August 22nd, 1885, complaining of thirst and of passing a great quantity of water.

Condition on Admission.—She was poorly nourished, but her physical signs were normal. The urine was 136 oz., sp. gr. 1039, acid, clear, a cloud of albumen; sugar 7·6 per cent.; urea 0·65 per cent.; acetone reaction and ferric chloride reaction both well marked. Temp. 98·8°, Pulse 108, Resp. 24. She was placed on house diet and ordered a simple enema.

Progress of Case.—On the evening of the 23rd she had an enema, and felt faint after it. She vomited just before midnight, and again at 4 a.m. At 5 o'clock she cried, and complained of great pain in the small of her back. The matter vomited gave a distinct acetone reaction with the nitro-prusside test. Hot fomentations were applied, and at 8 a.m. she had a grain of extract of opium, and a second was given at 10 o'clock. At the visit at 10 a.m. she was still complaining of the pain; her respiration was panting, 32; pulse 132; temp. 98·5°; bowels had been moved freely; tongue red and dry. She had passed no water since 4 a.m., so at 11 a catheter was introduced and 10 oz. of urine drawn off, which was pale, clear, sp. gr. 1040, acid, with a cloud of albumen, sugar 5·9 per cent., urea 0·9 per cent., acetone and ferric chloride reactions well marked. She was becoming unconscious, and by 11.30 was lying on her back comatose, with contracted pupils. At 11.30 thirty-six oz. of sulphate of soda solution were injected into the veins. She improved during the operation, the pulse became stronger, she was less drowsy, and she complained of the tightness of the bandage which was afterwards applied to her arm. At 2 p.m. she could not swallow, so three-quarters of a pint of water were administered by the bowel. At 3.30 the pulse was 165, very feeble, eyes turned up, respiration sighing. At 4 p.m. a second enema of about a pint of acidulated water was given. The pulse improved a little and respiration was quieter. At 5.30 16 oz. of urine were drawn off, which did not differ materially, except that it was of sp. gr. 1031 and contained only 4 per cent. of sugar and 0·3 per cent. of urea. An enema of acidulated water was given every hour; but at 7.30 p.m. the noisy deep breathing began again, and the enemata were discontinued as they were not retained. At 7 p.m. the temp. was 103°, and at 9 p.m. it had risen to 103·5°. Resp. 22 ; Pulse 169. No other change took place. She died quietly, without any convulsion, at 11.10 p.m.

At the autopsy the body was covered with a thick layer of fat. There was no peculiar smell on opening the body. The blood appeared normal. All the internal organs were congested but healthy, though the brain presented some excess of cerebro-spinal fluid.

DIAGNOSIS.—I do not think anyone can find a difficulty in diagnosing this condition, when he has once grasped its salient features. The peculiar dyspnœa, which is so characteristic, at once arrests attention. This is regarded by Senator as a pathognomonic symptom, and our present knowledge justifies his view. But it is not so easy, and, if possible it is of greater practical importance to recognise those prodromata, which foreshadow the approach of these formidable and almost invariably fatal symptoms.

The statistics of Guy's and the London Hospital, shew very clearly the greater frequency of this mode of termination of diabetes in young persons, so that youth *per se* appears to be a predisposing cause. Sex has no special influence that has been yet made out, nor does such an influence seem probable. Condition of life also appears unimportant.

Constipation has been a marked feature in several of my cases, and, as proved by the *post mortem* examination of the intestines, has existed to a far greater extent than had been suspected during life, so that I rate it highly in the list of predisposing causes. That the intestines are the seat of numerous fermentative and putrefactive processes is now well known, and it is easy to understand that constipation acts unfavourably in two ways : 1, by diminishing elimination of effete matter by one of its ordinary and most important channels, and 2, by affording time for the development of fermentative processes giving rise to the formation of toxic substances which may be absorbed into the blood.

Among the predisposing causes we may include a great increase of the acids of the blood. Minkowski's case, and another of the same kind, in which large quantities of acid were present in the urine, died of coma, so that we have in this an indication which may serve as a warning.

The determining causes are very numerous. Almost anything which is capable of producing a depressing influence, exertion, great emotion, anger, excitement, fear, fatigue, a local affection such as a strangulated hernia (Dixon Mann), a carbuncle or dental abscess, may act as the immediate agent in bringing about this train of symptoms.

It has been suggested that sudden change from ordinary diet to richly albuminous food may sometimes act as a determining factor in the production of this condition. I think the possibility of such an influence must be admitted, yet no such factor has been present in most of the cases which have come under my notice, and in one instance the use of an exclusive diet of skim milk seemed to be the immediate cause of the breakdown.

In the case of F. R. W., I have no doubt that the visit he made to his mother led to his death, by causing fatigue.

It is only too true that the life of a diabetic hangs by a thread, and we cannot be too careful or too persistent in our warnings. Above all we should never allow cases of advanced diabetes to travel; too many of those who go to Vichy and other resorts in the hope of finding a cure, die very soon after their arrival. This fact was insisted on by Prout, and is well illustrated by examples related by the physicians at the French and German Spas frequented by diabetics.

The first warning of the supervention of these symptoms has been, in two cases of my own, the discovery of a very marked ferric chloride reaction in the urine. In the case of F. R. W., it is very remarkable that this reaction, which had been repeatedly but not regularly looked for, was never noted till just before his death. I think, therefore, that restricting this statement to cases of diabetes, the sudden occurrence of this reaction should always serve as a warning. Stadelmann attaches much importance to the increase of the ammonia in the urine.

Rapid disappearance of sugar from the urine has been stated by some writers to have preceded the attack, but this is exceptional, as is also any marked alteration in the quantity of urine. Of more certain value than any of the foregoing is *epigastric pain*. This often precedes by some hours any other symptom. Next to this, in chronological order, is *rapidity and feebleness of the pulse*. But most significant of all is the *altered respiration*. Inspiration and expiration become prolonged, and the respiratory rhythm is quickened, but air enters freely into the chest; the lips and cheeks retain their former colour. Sooner or later the patient becomes *drowsy*, but it is easy to awaken him from this somnolent state. Convulsions may appear, usually towards the termination of the case; and before death cyanosis is often present, rather, I suspect, from failure of the circulation than from any difficulty in the oxygenation of the blood in the lungs.

A very striking symptom, when present, is the peculiar odour of the breath; but this is so far from being a constant phenomenon that Küssmaul does not mention it in his description, and I believe it to be present only in a minority of cases. It has been described in very varying ways as *sour*, like vinegar, sour beer, acetone; and *sweet*, like apples, or hay. It can hardly be that these different odours depend on the same substance.

20th. Pain at epigastrium returned. The pain on micturition had disappeared since admission.

Ordered, *Tr. opii* ℔xv.; *sodii salicylatis* gr. xv.; *inf. calumbæ* ʒj.; thrice daily.

31st. He went home at his own request, and we heard that he died very soon after.

The following table shews the effect of treatment.

Date.	Urine.	Sp. Gr.	Urea.	Sugar.	Wt. of Patient.
Dec. 13th.	188 oz.	1032	1·27 p.c.	4·7 p.c.	8st. 2lbs.
„ 16th.	136 oz.	1030	1·5 p.c.	4·0 p.c.	
„ 19th.	160 oz.	1028	1·1 p.c.	3·7 p.c.	8st. 1½lbs.
„ 25th.	168 oz.	1035	1·1 p.c.	5·5 p.c.	8st. 2¼lbs.

CASE 34.—*Diabetes Mellitus—polyuria—thirst—wasting —threatening coma averted by spontaneous diarrhœa.*

William S., aged twenty-one, engine cleaner, was admitted into hospital on March 27th, 1887, complaining of thirst, loss of flesh, and polyuria.

History.—His illness began six weeks before with thirst. He had a "low fever" three years ago, and formerly suffered from some kind of urticaria when working as a brass filer. He had always been temperate, and had a comfortable home. Father and mother and all his brothers and sisters alive and well; no history of diabetes in his family.

Condition on Admission.—He was rather deaf, especially in the left ear. He was wasted, and looked ill. Except some dulness and deficient breathing at the left apex posteriorly, the physical signs were normal. Urine 250 oz., sp. gr. 1037, acid, pale straw colour, no deposit, 0·8 per cent. of urea, 6·2 per cent. of sugar, a faint trace of albumen, a few squamous epithelial cells and crystals of oxalate of lime.

Progress of Case.—On March 31st he was dieted and treated with alkalies. His urine fell to half the quantity.

On April 3rd he complained of pains in head and abdomen, seemed drowsy, and had sighing breathing. His bowels, which had been acting once daily, were moved spontaneously four times, and the condition passed off.

He left at his own request on April 18th, and we heard in July that he had died.

TREATMENT.—Prevention is proverbially better than cure, but I need not repeat what has been already said respecting the predisposing and exciting causes, and will assume that all possible means will be taken to guard against them. In addition, two plans of treatment by drugs suggest themselves as theoretical prophylactic measures. The first is to check the fermentative processes by means of antiseptic or antizymotic drugs. For this purpose Foster recommended thymol, but I am not aware that he has ever given it a trial. Salicylate of soda has been used very extensively in diabetes, but in spite of it these symptoms have supervened. Nevertheless, in selecting new remedies we should bear in mind that this action of the salicylate is one among other reasons for its employment.

The other plan is the use of alkalies. This mode of treatment has received the sanction of the highest authorities, yet it is at present in some danger of falling into disfavour. A high degree of acidity of the urine should be regarded as an indication for its use, Vichy water being as good as any other mode of administration.

Is there any remedy which can arrest the attack, when it has commenced ?

Frerichs regards all therapeutic measures hitherto suggested as useless. Injections of ether, camphor, and similar stimulants have been tried without advantage.

Küssmaul tried transfusion of blood with only temporary results, and inhalation of oxygen was also without effect. Bence Jones employed peroxide of hydrogen and stimulants with no better result. Hilton Fagge and Taylor have injected a weak solution of phosphate of soda and sodium chloride into the veins, in one case with benefit for some hours, but after a dose of codeia the coma returned and proved fatal. In another case no result followed.

Rest in bed should always be insisted on when any threatening symptoms appear. In the early stages I would recommend energetic means to obtain an evacuation of the bowels, combined with the administration of stimulants and alkalies by the rectum. Should these fail, we may always try the effect of intra-venous injection of a saline solution.

These three cases shew the results of the use of various remedies, intra-venous injections and intra-peritoneal injections of saline solutions, hypodermic injections of strychnine, inhalations of oxygen, etc.

CASE 35.—*Diabetes Mellitus—polyuria—thirst—emaciation—rheumatic antecedents—unfavourable effect of skim milk—death by coma—use of strychnine.*

Arthur P., forty-three, nailer, was admitted into the General Hospital on October 20th, 1886, complaining of thirst, weakness, and of passing a large quantity of water —he thought as much as six or seven quarts a day.

History.—His illness began six months ago with pains in his legs, which his doctor called rheumatism. A month later it was discovered that he had diabetes, but no change was made in his diet. He continued to get worse; and having been to Rhyl without benefit he came to the hospital. His previous health had been good, except for rheumatic pains in his hands. His father and mother were alive, as well as all his brothers and sisters, and, except that his mother suffered from rheumatism, no pathological incident could be discovered.

Condition on Admission.—He was a small, emaciated man; height 5 ft. 3 in.; weight 8 st. ¼lb. He used to weigh 10 st. Temp. 98°; Pulse 72; Resp. 14. Teeth irregular and decayed, tongue red and beefy, flatulent eructations after food, bowels regular, liver and spleen normal. Complains of giddiness, and is short of breath on exertion. Heart's sounds normal; no cough; some impaired resonance, harsh breathing, prolonged expiration, and slight crepitation at the right apex. Urine 264 oz., sp. gr. 1035, acid, clear, urea 0·7 per cent., very faint cloud of albumen, sugar 6·9 per cent.

Progress of Case.—On Oct. 22nd he was put on six pints of skim milk daily.

Oct. 23rd. He grumbled about his diet.

Oct. 24th. Complained of feeling low; bowels open; tongue still raw and beefy. Urine 216 oz., sp. gr. 1033, acid, pale straw colour, white flocculent deposit, urea 0·9 per cent., a cloud of albumen, sugar 6·25 per cent., a trace of blood, and a few red and white blood corpuscles seen with the microscope. At 7.30 p.m. he complained again of feeling low, and of a pain in his bowels which he attributed to the milk.

Oct. 25th. In the morning he made the same complaint, and said he was cold. His pulse was feeble, and his tongue dry, brown, and cracked. At 10 a.m. he was drowsy, and his respiration was rather deep. At 11.30 a.m. Pulse 160; Resp. 20; there was pain in the stomach and bowels; no urine passed since 8 a.m. He was ordered an enema to empty the bowels, a bottle of Vichy water to be taken with lemon juice in the course of the day, an alkaline injection (*Sodii bicarb.* ʒi., ad. *aq.* Oss.) to be given by the bowel as soon as the rectum had been cleared out, and the following mixture every hour: ℞. *Etheris sulphuris* ♏xx; *Potassii bicarbonatis* gr.x; *Aquam ad.* ʒi., M. In the course of the day the rectum was well cleared out and scybalæ removed, but he got worse, and at 8.45 p.m. I saw him again. Temp. 99; Resp. 24. He lay apparently asleep, but could not be roused. His breathing was very deep, but there was no special odour about him. His urine gave a well-marked acetone reaction. He was ordered liq. strychninæ ♏ v. *sub cute* every fifteen minutes until muscular twitchings were produced. After three injections he spoke and drank some Vichy water. At 9.25 he had another alkaline enema, containing one ounce of bicarbonate of soda. After this the strychnine injections were given irregularly at 10.10, 10.40, 11.10, 11.55, 1.50 a.m., and then not till 9 a.m.

26th. At 9.30 a.m. I saw him again. No physiological symptoms had been produced by the strychnine, so ten minims were ordered to be injected every half-hour. Pulse 128; Resp. 26; Temp. 99. He was decidedly better, answered when spoken to, took beef tea very well. The knee jerks were absent. The strychnine was again neglected; after I left it was given only every hour. His temperature had begun to rise. At 11.30 it was 102°; at 12.30, 103°; by 2 p.m. he was comatose again. Three ounces of urine were drawn off from his bladder; it was acid, sp. gr. 1023, and free from albumen and sugar. At 5 p.m. his temperature was 105°, and at 6 p.m. reached 106°. At 6.30 he was plainly dying; the face was a little drawn to the left, and the right pupil was larger than the left. He died at 6.40 p.m. without any other change taking place.

AUTOPSY.—A spare subject, somewhat emaciated, faint sweetish ethereal odour emanating from cadaver, p.m. rigidity slight, hypostasis chiefly posterior.

Brain Membranes.—Excess of serous fluid in subarachnoid space and in ventricles; arachnoid very œdematous, cerebral convolutions well defined, no change to naked eye.

The corpora striata and optic thalami shewed no change in section except the 'pin holes' described by Dickinson; these were certainly evident and scattered chiefly in the white substance of the internal and external capsule. Medulla: no change recognisable by naked eye; it was kept for microscopic investigation. Pons normal.

Heart.—8 oz., normal.

Lungs.—*Left* upper lobe solid with caseous pneumonia; the interlobular peribronchial fibrous connective tissue was likewise thickened. Pleura over upper lobe thickened and adherent; lower lobe congested and œdematous. *Right Lung*: one or two small foci of caseous pneumonia at apex.

Liver.—59 oz., surface of a dark reddish brown colour. Section: the organ was generally hyperæmic, of a dark red colour, but scattered here and there in an irregular fashion were small yellowish patches corresponding to two or more lobules, where the liver cells microscopically were very fatty.

Kidneys.—10 oz., capsules stripped easily, both very dark from hyperæmia, the streaky and punctiform distribution of which was very obvious; the glomeruli appeared as minute red points, slightly elevated above the surface; the medullary cones were of deeper hue.

There was a faint sweetish ethereal odour emanating from the viscera.

No mention of the condition of the pancreas or semilunar ganglia.

CASE 36.—*Diabetes Mellitus—thirst—polyuria—wasting —family history of phthisis—acetonuria—coma—failure of strychnine injections—oxygen inhalations—and intraperitoneal alkaline injections—death.*

Mary Ann M., aged thirty, screw-maker, was admitted into the General Hospital on September 10th, 1889, complaining of great and increasing weakness, excessive thirst, and increased secretion of urine. These symptoms came on gradually four years ago, accompanied by low feelings, fainting fits, sickness, and shortness of breath. Her hunger was at that time so great that she could not get enough to satisfy herself, and she began to lose flesh, to sweat a great deal at night, and to suffer from a troublesome itching of the skin, which was worse at night. During the whole time her bowels had been very constipated, having been open only once or twice a week. Micturition had been very frequent and attended by smarting. For the last month there had been some pain and discomfort after food; and her ankles had got swollen after she had been standing. Three days before admission she had felt so low that she had gone to bed, and had remained there ever since.

Family History.—Father alive, but suffering from phthisis. Mother and seven brothers alive and well. Seven other children died young. Patient had been married ten years, but had had no children.

Previous History.—Could remember no other illness.

State on Admission.—Patient was rather a spare woman with a flushed face. Temp. 99·5°; Pulse 88. Weighed 7 st. 5 lbs.; former weight 11 st. 4 lbs. There was a peculiar sweet smell emanating from her.

Alimentary System.—Teeth decayed posteriorly; gums bled readily; subject to toothache; teeth were loose, but had got firmer; a nasty taste in the mouth, worse in the morning; mouth dry; great thirst; appetite voracious. Tongue large, dry, cracked transversely, and covered with thin dirty fur on dorsum. Some pain and discomfort after food. Stomach dilated, reaching nearly to umbilicus, with splashing sound on palpation. Liver dulness 4 inches in V. M. L. Splenic dulness not increased.

Circulatory System.—No pain or palpitation; some dyspnœa. Area of cardiac dulness not increased; apex beat in 4th I. S., inside the V.M.L. Sounds clear; pulmonary second accentuated. Pulse 96, not easily compressible.

Respiratory System.—No cough, pain, or dyspnœa; percussion note deficient in resonance at right base.

Urinary system.—Urine 90 oz., sp. gr. 1034, acid, pale amber; urea 1·1 per cent., a very faint haze of albumen; sugar 5·4 per cent.

Treatment.—Diet: Mutton, meat jelly, milk two pints, cold meat two portions, Vichy water with lemon juice. Medicine: *Extracti belladonnæ* gr. ¼; *extracti aloes* gr. iij *euonymini* gr. j.; in pill at bedtime.

July 15th. Complained of feeling very sick; ordered an enema, which brought away a motion containing a lot of hard lumps. Urine 1025, highly acid; contained a haze of albumen, gave a deep Burgundy red coloration with ferric chloride, and a moderate rose-violet coloration with nitro-prusside of sodium and ammonia. At *noon*, face flushed, epigastric pain, pulse 132, tongue dry, respiration 24, sighing. Ordered, *liquoris strychninæ* ♏v., *sub cute quartis horis*; and to continue the Vichy water without the lemon juice. Later in the day her extremities got very cold.

July 16th. 4 a.m., was conscious. 8.30 a.m., pulse scarcely perceptible; hands not cold; unconscious; passed water in bed. Respirations irregular, but not Cheyne-Stokes' type. Pulse 128; Resp., 32. 10 a.m. extremities cold. 10.30. passing plenty of water in bed; swallowing difficult. Strychnine to be injected every half-hour. This was done regularly until 4.35 p.m. inclusive, except at 3.30. 11.30. Oxygen was administered by means of a nitrous oxide gas apparatus. During the inhalation the pulse became *weaker*. The whole of the oxygen contained in a cylinder was given, but without the slightest benefit. 3.30 p.m. A solution of bicarbonate of soda (2 oz. of the salt dissolved in 2 pints of water at 100° F.) was slowly introduced into the peritoneal cavity. 4.15 p.m. No result had followed the first injection, so two pints more were introduced. The patient seemed better. She took some milk and beef tea; and just as there seemed a ray of hope that at least temporary improvement had been secured, she quite suddenly looked worse, there was a slight spasm of muscles of the face and left arm, and she died in spite of the prompt injection of a syringeful of ether.

Autopsy.—A spare but fairly nourished woman. Abdomen distended, and on opening it a large quantity of yellowish tinged fluid escaped. The retro-peritoneal cellular tissue was much infiltrated with fluid, also the deep fascia and inter-muscular connective tissue of the lower part of right side of thorax.

Brain.—46 oz., anæmic, nothing abnormal recognised with naked eye. Pons and medulla preserved.

Heart.—8¼ oz., L.V. was firmly contracted and appeared thicker than normally, measuring nearly $\frac{1}{10}$ inch. No valve lesions, heart otherwise quite healthy.

Lungs.—2 lbs., both congested; hypostasis of the bases.

Liver.—65 oz., much engorged with blood. Veins dilated. Colour of parenchyma a dark reddish brown, but mottled here and there with yellowish patches.

Kidneys.—15 oz., large, capsules tense, slightly sticky. On section both looked fatty. Cortex increased and variegated by deep red striæ alternating with an opaque yellowish grey labyrinth, both considerably congested.

Pancreas 2¾ oz. small, gland substance looks shrunken, consistence generally tough and fibroid, cuts with perceptible resistance to knife's edge.

Semilunar ganglia.—Looked small, weighed together 57 grains.

Stomach.—Much distended, partly with a dark brown grumous fluid; mucous membrane pale but remarkably mammillated.

Case 37.—*Diabetes Mellitus—thirst—hunger—wasting—polyuria—heredity—vertigo—diarrhœa—acetonuria—sickness—coma—no ferric chloride reaction—death—autopsy.*

Amelia P., aged thirty-seven, domestic servant, was admitted into the General Hospital October 23rd, 1889, complaining of hunger, thirst, wasting, and polyuria. These symptoms had existed for ten months.

Previous History.—She could remember no previous illness or injury; her health had been good, but she was subject to frontal headaches. Six years ago she had an "abscess" over the left eye, and the pain had been worse in that spot ever since. She had worn spectacles since she was twelve years of age, and had supported herself since she was thirteen. The work had been hard, and she had had a good deal of worry, but had always plenty to eat. She had been accustomed to take very little beer, about one glass for supper.

Family History.—Father died of bronchitis aged fifty-three. Mother died of diabetes aged thirty-seven; in the latter part of her illness she had many "fits." One brother died in the General Hospital after an accident; two others were living in good health. She did not know of any insanity or phthisis or any other case of diabetes in the family.

State on Admission.—A small slightly developed woman. Weighed 6 st. 10 lbs., but used to weigh 8 stone or more. She looked wasted. Her face was a peculiar brownish yellow colour, which she said was a recent change; her cheeks were flushed and her lips red. Eleven days before admission she had a fit of giddiness in which she fell down and abraded the skin on the bridge of her nose. She believed that she lost consciousness completely. Temp. 97°, Pulse 84, Resp. 18.

Alimentary System.—She was very hungry, craving food every two hours, and her thirst was also great. Bowels regular. Tongue pretty clean, furrowed, dry. Teeth defective, decaying. Flatulence after eating, but no pain or tenderness. She had had one or two severe attacks of diarrhœa. Abdomen not distended. Liver dulness in V. M. L. $3\frac{1}{4}$ inches. Splenic dulness in M. A. L. $1\frac{1}{4}$ inches.

Circulatory System.—Heart's area not increased. Apex in 4th I. S. in V. M. L. Sounds normal. Pulse 84, regular, small and soft.

Respiratory System.—Chest movements normal, no cough. Percussion and breath sounds normal.

Genito-urinary System.—She had not menstruated since November (12 months); no vaginal discharge. Frequent micturition. Urine 244 oz., sp. gr. 1035, acid, pale yellow; urea 0·8 per cent., no albumen; sugar 5·7 per cent., acetone present, no deposit.

Treatment.—Diet: meat, beef jelly, fish, chicken, two eggs, no bread or vegetables, toast-water, Vichy water and lemon juice. Medicine: *Ex. Opii* gr. j.; twice daily.

Progress of Case.—Oct. 26th. Between 3 and 4 a.m. she had a severe attack of diarrhœa, followed by a feeling of sickness, and was drowsy and heavy all day. Complained of feeling hot and feverish, but temperature was 98·5°.

Oct. 27th. Complained of sickness and a sinking feeling; ordered half a pint of milk, an effervescing mixture, and some gluten bread.

Oct. 28th. She would not eat the meat alone, so was allowed green vegetables, and as no gluten bread had arrived, two small slices of toast were allowed.

Oct. 29th. Sick again in the morning. Respiration sighing at times towards evening.

Oct. 30th. Was restless and moaning all night. Between twelve and one she was delirious, screaming and trying to get out of bed. Respiration was sighing, 20. Pulse 96. Hands and feet cold; not conscious. No water had been passed since 9 p.m. 11 a.m. she was lying quietly, respiration sighing, face pale, extremities cold; no peculiar odour of breath. Urine, drawn off by catheter, 7 oz., pale greenish, clear, acid, sp. gr. 1017, a cloud of albumen, loaded with sugar, faint acetone reaction with nitro-prusside of sodium, *no reaction with ferric chloride.* 11.45 almost pulseless; a syringeful of ether was injected subcutaneously. 12.15 two pints of the following solution, at the temperature of the body, were injected into the peritoneal cavity:—*Sodii sulphatis;*

sodii bicarbonatis āā ʒj.; aquæ bullientis ʒ viij.; solve; adde aquæ destillatæ, Oj.

This was followed by no improvement, and she died at 2.15. p.m. the same day.

AUTOPSY.—Height, 60 inches; circumference of chest, 27 inches; rigidity well marked; puncture seen in linea alba midway between umbilicus and pubes.

Thorax.—Heart 7 oz.; large milk spot over right ventricle; much fat over right ventricle externally. Left ventricle, heart substance slightly fatty; mitral valve normal; aortic valve normal and competent, commencing atheroma at the beginning of the aortic arch. Right ventricle contained very dark p.m. clot; tricuspid and pulmonary valves competent and normal; auricles normal. Left lung: 8 oz.; substance healthy; no bronchitis; was bound down to chest wall by few adhesions. Right lung: 10 oz.; normal.

Abdomen.—Liver: anterior surface adherent to the diaphragm. On section, consistence rather tough. Groups of lobules appeared translucent surrounded by yellow streaks. Gall-bladder, collapsed; on squeezing it yellowish-green bile exuded. Right kidney: 6 oz.; capsule slightly adherent, taking away a little kidney substance; cortex pale yellow; interpyramidal cortex the same, streaked with translucent yellow lines; malpighian bodies very prominent through the whole cortex, arranged in rows; pyramids lightish red, streaked with yellow. Left kidney: 6½ oz.; same condition as right.

Brain.—3 lbs.; subarachnoid fluid over both hemispheres. Section shewed brain substance slightly congested. Right semilunar ganglion somewhat larger than left.

CONCLUSIONS.—The following conclusions summarise briefly the most important facts contained in this long chapter:—

1. Diabetic coma is specially apt to supervene in acute cases in young persons.

2. Diabetic patients and their friends should be warned of the danger of constipation, muscular exertion, nervous excitement, and cold, as probably predisposing causes of death by coma.

3. The discovery of the ferric chloride reaction in the urine should be taken as a warning against the premonitory symptoms of coma.

4. Deep respiration, rapid pulse, and abdominal pain are the earliest premonitory signs of this condition.

5. Cyanosis may be absent in spite of the dyspnœa, and may appear only just before death.

6. Convulsive seizures are not an uncommon occurrence just before death.

7. The temperature is usually normal or sub-normal, but may be considerably raised.

8. Diabetic coma, with all its classical symptoms, occurs independently of any excess of fat in the blood, and the pathological value of lipæmia, when present, is yet undetermined.

9. The toxæmic theory, *e.g.*, poisoning by acetone or some nearly allied substance or substances, affords the best explanation of this remarkable group of symptoms.

10. Recovery is possible from the prodromal symptoms, and even from some degree of drowsiness, but from actual coma it is at least very rare.

11. Great benefit may ensue in the early stages from speedy evacuation of the bowels by a brisk purgative. Treatment in the later stages seems always unavailing.

BIBLIOGRAPHY.

BINZ. Pathology of Diabetic Coma. "Cong. für inn. Med. Wiesb.," 1886.

BOND and WINDLE. Diabetes terminating by Coma. "Brit. Med. Jour.," 1883, I., p. 908.

BUHL. Ueber diabetisches Koma. "Zeit. f. Biol.," Bd. XVI., 1880.

BULL (G. C.). A preliminary Report on the presence of Acetone in the Urine. "Birm. Med. Rev.," Vol. XVII., 1885, p. 211.

CANTANI (A.). Der Diabetes mellitus. Berlin, 1887.

CHARCOT (J. M.). The Causation of Diabetes. "Journ. de Méd.," April, 1883.

CYR. De la mort subite ou très rapide dans le Diabète. "Arch. Gén. de Méd.," 1877, II. and 1878, I.

DRESCHFELD (quoted by GAMGEE, A.). A Text-book of Physiological Chemistry of the Animal Body. London, 1880.

DRESCHFELD (J.). The Bradshawe Lecture on Diabetic Coma. "The Lancet," 1886, II., p. 369.

EBSTEIN (W.). Ueber Drusen-Epithelnekrosen beim Diabetes mellitus, mit besonderer Berücksichtigung des diabetischen Coma. "Deutsches Archiv. f. klin. Med.," Bd. XXVIII., 1881, p. 143.

FAGGE (C. HILTON). *Op. cit.*

FLEISCHER (R.). Beitrag zur Chemie des diabetischen Harns. "Deut. med. Woch.," 1879, No. 18.

FOSTER (B.). Diabetic Coma—Acetonæmia. "Brit. Med. Jour.," 1878, I., p. 78.

FOSTER and SAUNDBY. Diabetic Coma. "Birm. Med. Rev.," Vol. XIII., 1883, p. 1.

FRERICHS. Plötzlicher Tod und Coma bei Diabetes. "Zeit. für klin. Med.," Bd. VI., Heft 1. 1883.

HESSE. Coma benefited by intra-venous injection of 4 per cent. solution of Carbonate of Soda. "Berlin. klin. Woch.," No. 19, 1888.

HOPPE-SEYLER (G.). Ueber das Auftreten acetonbildender Substanz im Urin nach Schwefelsäurevergiftung. "Zeit. f. klin. Med.," Bd. VI., p. 478.

JONES (H. BENCE). On Animal Chemistry, in its application to Stomach and Renal Diseases. London, 1850.

KÜSSMAUL. Zur Lehre vom Diabetes mellitus. "Deut. Archiv. f. klin. Med.," 1874.

LATHAM (P. W.). Pathological relation between Diabetes, Gout, and Rheumatism. "Brit. Med. Jour.," 1885, II., p. 1053.

LE NOBEL (C.). Ueber einige neue chemische Eigenschaften des Acetone und verwandter Substanzen und deren Benutzung zur Lösung der Acetonuriefrage. "Archiv. f. exp. Path.," Bd. XVIII., Heft. 1 and 2.

LÉPINE (R.). Sur la pathogénie et le traitement du coma diabétique. "Revue de Méd.," Vol. VII., 1887, p. 224.

LITTEN (M.), Ueber einen eigenartigen Symptomencomplex in Folge von Selbstinfection bei dyspeptischen Zuständen (Coma diabeticum). "Zeit. f. klin. Med.," Bd. VII.

MACKENZIE (S.). The Pathology of Diabetes. "Brit. Med. Jour.," 1883, I., p. 655.

MINKOWSKI (O.). Ueber das Vorkommen von Oxybuttersäure im Harn bei Diabetes mellitus. Ein Beitrag zur Lehre von Coma diabeticum. "Arch. f. exp. Path.," Bd. XVIII., 1884.

MINOT (F.). Sudden Death in Diabetes. "Boston Med. and Surg. Jour.," Jan. 19th, 1888.

MOTT (F. W.). Four cases of Diabetic Coma. "The Practitioner," Vol. XL., p. 420.

PENZOLDT (F.). Ueber den diagnostichen Werth d. Harnreaktion mit Diazobenzolsulphosäure u. über deren Anwendung zum Nachweis von Traubenzucker. "Berlin. klin. Woch.," Bd. XX., 1883, p. 14.

PROUT (W.), On the Nature and Treatment of Stomach and Renal Diseases, Fifth edition. London, 1848.

QUINCKE (H.). Ueber Coma diabeticum. "Berl. klin. Woch.," Bd. XVII., 1880.

RICKARDS (E.). A case of Diabetes, terminating suddenly with milky blood. "Birm. Med. Rev.," Vol. XI., 1882, p. 271.

ROBERTS (PRESCOTT). A case of Diabetic Coma. "Brit. Med. Jour.," 1882, II., p. 1038.

ROBERTS (W.). Urinary and Renal Diseases. Fourth edition. London, 1885.

SALKOWSKI and LEUBE. Die Lehre vom Harn. Berlin, 1882.

SALOMON and BRIEGER (*vide* FRERICHS). "Plötzlichen Tod und Coma."

SANDERS and HAMILTON. Lipæmia and Fat Embolism in the Fatal Dyspnœa and Coma of Diabetes. "Edin. Med. Jour.," Vol. XXV., 1879, p. 47.

SAUNDBY (R.). Diabetic Coma. "Birm. Med. Rev.," Vol. X., 1881, p. 193.

———— and BARLING. Fat Embolism. "Jour. of Anat. and Phys.," Vol. XVI., 1882, p. 515.

————. Sudden Death in Diabetes. "Birm. Med. Rev.," Vol. XV., 1884, p. 148.

————. Küssmaul's Coma. Ibid., Vol. XVII., 1885, p. 49.

SCHMITZ (R. W.). On a frequent and noteworthy Complication of Diabetes Mellitus. "Berlin. klin. Woch.," Jan. 31st, 1876.

————. Zur Behandlung des Coma diabeticum. "Berlin. klin. Woch.," 1890.

SENATOR. "Ziemssen's Cyclopædia." Vol. XVI., p. 916.

————. On self-infection by the products of abnormal processes of decomposition within the body, and the resulting (Dyscrasia) Coma (Küssmaul's Diabetic Coma). "Zeit. f. klin. Med.," Band VII., Heft 3.

STADELMANN. Coma diabeticum und seiner Behandlung. "Deut. med. Woch.," 1889, No. 46. "Deut. Archiv. f. klin. Med.," Bd. LXXXVI., Heft 6.

STOCKVIS and HOFFMANN. Zur Pathologie und Therapie des Diabetes mellitus. "Cong. für inn. Med. Wiesb.," 1886.

TAYLOR (F.). On the fatal termination of Diabetes: with especial reference to the death by Coma. "Guy's Hosp. Rep.," 3rd S., Vol. XXV., 1881.

VON JAKSCH (R.). Ueber Coma carcinomatosum. "Wien. med. Woch.," 1888, Nos. 16 and 17.

————————. On the red colour produced in Diabetic Urine on the addition of Chloride of Iron. "Zeit. f. Heilk.," Bd. III.

————————. On the presence of Aceto-acetic Acid in the Urine. "Zeit. f. phys. Chem.," Bd. VIII., Heft 6, 1883.

WALTER. Untersuchungen über die Wirkung der Säuren auf den thierischen Organismus. "Arch. f. exp. Path.," Bd. VII.

WINDLE (B. C. A.). On the Ferric Chloride Reaction in Urine. "Liverpool Med. Chir. Jour.," Vol. IV., 1884, p. 84.

Chapter VI.
TREATMENT.

Before we can estimate properly the effects of any plan of treatment, it is above all things essential that we should know the natural course of the ailment we are about to treat. There is in many diseases a spontaneous tendency to cure after the lapse of a certain period of time, and in some this tendency affords most providential assistance to an otherwise inefficient pharmacopœia. Modern criticism has exposed many of these fallacies. No one nowadays regards the crisis in pneumonia as the happy outcome of successful treatment. But there is a fascination in the belief that we are effecting a cure which makes this sort of delusion very hard to get rid of.

Hence it probably arises that a positive statement, though it be the simple assertion of an unknown and irresponsible person, is copied and recopied by the medical press all over the world, and survives all contemporary paragraphs on other subjects, while an account of the unsuccessful trial of a well puffed drug falls still-born from the press.

In diabetes there are two well-marked types, which differ essentially in their amenability to treatment. Although this is not by any means a new fact, it has not been accepted as widely as it should be.

The distinction to be drawn is between diabetes in people under forty-five and in those over that age. This discrimination is needed, because the diabetes in elderly people is frequently curable, and in most cases is controlled by the careful avoidance of saccharine and starchy articles of diet.

Diet.—The following two cases are examples of the control of this form of diabetes by proper diet.

CASE 38.—Josiah W——, fifty-six, a big, stout man, came to the out-patient department of the General Hospital complaining of "sciatica." His right leg was swollen and painful. As neuralgic pain in the leg is often associated with diabetes, I had his urine examined, and it was found to be loaded with sugar. He told us that he had suffered much from thirst, had lost weight, and was passing a great deal of water. On admission he was placed on a strict diabetic diet, with an alkaline mixture and imperial drink made with saccharin. The next day his urine was free from sugar. On being allowed to resume ordinary diet, his urine was found to contain 2·08 per cent. of sugar; on restricting his diet again the sugar disappeared, but was detected later on, and the cause traced to the surreptitious use of bread supplied him by a fellow-patient. This was prevented, and his urine remained free from sugar so long as the diet-regulations were enforced.

CASE 39.—Elizabeth G——, sixty-five, was admitted for pain shooting down the right leg. Her urine was loaded with sugar, but on strict diet this entirely disappeared, although at first there was a little difficulty owing to the fact that she had a packet of sweets in her locker.

Even in the generally more obstinate type met with in younger people proper diet always produces a decided effect, and sometimes this is so very marked that no drug-treatment can be properly tested until the effect of dietetic restrictions has been fairly tried.

The following case is a very good example of what may be done by diet alone to reduce the amount of sugar in a tolerably mild specimen of the early type, and of the ineffectiveness of the arsenite of bromine treatment. It also shews that small doses of cod-liver oil do not cause any increase of glycosuria.

CASE 40.—George B——, edge-tool grinder, aged twenty-eight, was admitted into hospital on September 26th, complaining of weakness, thirst, and pain in the left side after food. He had had "rheumatism" just after Christmas, but was not confined to bed, the pain being chiefly in the legs and feet. His previous health was good, but he used to drink a good deal of beer. He had never met with any accident nor had any severe blow anywhere. He was a pale, delicate, fairly-nourished man of middle height, weighing 8st. 9½lbs. in his clothes. Expression of face anxious. Ptosis of the right eyelid, which he said had always drooped. Knee-jerk present; abdomen flat; bowels confined; liver and spleen normal; tongue clean : pulse 66, regular; respirations 24; temperature 98·5°; heart and lungs normal. He was put on ordinary diet and passed about 120 oz. of urine daily, sp. gr. 1042, containing 8 per cent. of sugar. On October 1st he began diabetic diet, viz., meat, green vegetables, and gluten bread. The quantity of urine at once dropped to 60 oz., sp. gr. 1032. He began cod-liver oil on the 3rd, and on the 6th his urine was 64 oz., sp. gr. 1028, containing 1 per cent. of sugar. He began on the 7th to take five minims of Clemens' solution of arsenite of bromine three times a day, as well as the oil. The quantity of urine remained the same, but the specific gravity and sugar rose, being on the 12th 1037, with 3·9 per cent. of sugar. The arsenite of bromine was then increased to ten minims and the oil stopped, and on the 19th the urine was 60 oz., sp. gr. 1038, containing 4 per cent. of sugar. The drug was then omitted, and on the 22nd the urine was 52 oz., sp. gr. 1033, containing ·68 per cent of sugar, and on the 25th it was 54 oz., sp. gr. 1036, with 1·9 per cent. of sugar. The only other treatment employed was a vapour bath twice a week, and an occasional laxative. He gained in the month's treatment three pounds in weight.

In the treatment of diabetes we should set before ourselves four main objects, in the following order :—

1. To relieve the urgent and distressing thirst.

2. To diminish the quantity of urinary water.

3. To restore the body-weight to the normal healthy standard.

4. To remove, if possible, all traces of sugar from the urine.

The first two are symptoms of which patients complain loudly. The thirst torments them day and night, and the constant necessity to pass water makes continuous sleep impossible. But while we must relieve these, the improvement of general nutrition and the recovery of lost flesh are the best indications of successful treatment. All patients should be weighed when they come first under treatment, and afterwards every week to mark progress. The fourth object, though a consummation devoutly to be wished, is often impossible and sometimes unwise to attempt, for many elderly diabetics perceptibly decline in health if we enforce those strict rules of diet, which nevertheless control their glycosuria most effectually.

The cases I have quoted prove quite sufficiently that all these objects may be attained in some cases by dietetic restrictions alone, but these require to be formulated.

The fundamental elements of the diet of a diabetic patient should be meat, green vegetables, and gluten bread. The meat should be chiefly fresh, hot or cold, roast, boiled, or stewed, but some portion may be salted or smoked if desired. The vegetables may be raw or cooked.

The patient should be encouraged to eat fats, fat bacon, cream, eggs, and if necessary cod-liver oil should be added.

There can be no doubt that a purely *animal diet* is the best, but it is difficult to get patients to keep to such a *régime* for long together.

Of all the bread substitutes, Pavy's *Almond Cakes* are theoretically the best. They are quite free from starch, dextrine and sugar, and they are pleasant to eat. But they are not very digestible, and they do not quite supply the kind of food which is required to take the place of bread. Probably no food free from starch can do so.

Gluten bread is not a very satisfactory article, as it always contains a considerable percentage of starch. The following tables give the analysis of the best specimens I have been able to obtain in England and France:—

	French.	English.
Moisture	9·30	9
Sugar	2·50	1
Fat	2·50	14·3
Insoluble Carbohydrates	33·28	27·4
Gluten	49·62	55·5
Ash	2·80	1·8
Samples dried at 212°:—		
Gluten	54·71	55·5
Insoluble Carbohydrates	86·60	27·4

Ordinary wheat flour contains about 63 per cent. of starch, so that a patient who eats half-a-pound of gluten bread, gets as much starch as one who contents himself with four ounces of ordinary bread.

The use of "Soya" bread has recently been advocated by Dujardin-Beaumetz. It is made from the meal of Soya Hispida, a bean like the haricot, a native of China and Japan, but now cultivated in Austria. Bread made from it is not unpalatable and far superior to gluten bread in appearance and taste. A report in the *Lancet* gives the following figures respecting its composition:—

Nitrogenous material	25·02 per cent.
Starch	2·72 per cent.
Mineral matter	4·0 per cent.

"Soya" bread, biscuits, and flour are now prepared by a London firm. Samples procured from this firm, and submitted to analysis, give the following figures:—

	Soya Bread.	Soya Biscuits	Soya Flour.
Moisture	44·1	8·8	10·6
Fatty matter	4·2	17·1	4·6
Mineral matter	3·4	5·0	5·7
Carbohydrates	23·3	46·1	45·6
Nitrogenous matter	25·0	23·0	33·5
	100·0	100·0	100·0

This result was certainly very unexpected, but if it is to be taken as correct, it shews that the Soya Bread contains about as much carbohydrates as the best gluten bread, over which it has no advantages at present in the matter of cheapness. My patients have objected very much to its taste, and it appears to exert a laxative effect on the bowels, which is not always desirable.

Armitage has recommended the use of semolina. This substance is stated by Hassall to be wheat gluten, or wheat deprived of its starch.

A recently-advertised article called "Florador" is said to have much the same composition. I procured samples for analysis with the following results:—

	Semolina.	Florador.
Moisture	12·2	11
Fatty matter	1·1	0·8
Mineral matter	0·7	0·25
Carbohydrates	74·5	75·55
Nitrogenous matter	11·5	12·40
	100·0	100·0

The analyst adds a note that wheat starch formed the bulk of the carbohydrates present.

The following table from de Nédats, quoted by Dujardin-Beaumetz, is of value as shewing the amount of starch in various vegetable substances :—

Rice	74·10 per cent.
Maize	65·90 per cent.
Wheat Flour	63·00 per cent.
Wheat Meal	59·60 per cent.
Rye Flour	59·84 per cent.
Millet	57·90 per cent.
Sarrazin (Buckwheat)	50·00 per cent.
Wheat Bread	42·70 per cent.
Oatmeal	39·10 per cent.
Peas	37·00 per cent.
Rye Bread	36·25 per cent.
Haricots	36·00 per cent.
Jerusalem Artichokes	16·60 per cent.
Potatoes	15·50 per cent.

In accordance with these figures Dujardin-Beaumetz allows his patients 100 grammes (7 ounces) of potatoes daily, as containing less starch than gluten bread.

Buckwheat flour is thought in America to possess certain advantages over wheat flour, and as the above table shews it contains considerably less starch; buckwheat cakes probably contain only about 33 per cent. of starch, and thus are superior to at least some of the gluten bread in the market.

O'Donnell recommends the following as a home-made substitute for bread: Six eggs are thoroughly beaten, and then a teaspoonful of baking powder or its chemical equivalent, and a quarter of a teaspoonful of salt are added, and again the eggs are beaten. This mixture poured into hot waffle irons, smeared with butter, is baked in a very hot oven. For variety, and to make these biscuits seem more like coarse bread, pulverized nuts (almonds) may be added. They may be eaten hot with butter and cheese, but will keep for a long time, and nobody would suspect that they were destitute of flour.

With respect to the influence of potatoes on the excretion of water and sugar, I made the following observations on two serious and since fatal cases of diabetes.

CASE 41.—H. G., aged thirty-six :—

Diet—Meat, greens, and gluten bread. Duration of observation—4 days. Body weight—6st. 13lbs. Average daily quantity of urine—135 ozs. Average daily percentage of sugar—8·6 per cent.

Diet—The same, with the addition of potatoes. Duration of observation—15 days. Body weight—7st. 4¼lbs. Average daily quantity of urine—132 ozs. Average daily percentage of sugar—6·4 per cent.

Diet—The same, without potatoes. Duration of observation—13 days. Body weight—7st. 4¼lbs. Average daily quantity of urine 136 ozs. Average daily percentage of sugar—6·8 per cent.

Diet—The same, with potatoes. Duration of observation—19 days. Body weight—7st. 2½lbs. Average daily quantity of urine—127 ozs. Average daily percentage of sugar—6·8 per cent.

The general result may be summarised in the following table:—

	With Potatoes.	Without Potatoes.
Urine	129·5 ounces	135·5 ounces
Sugar	6·6 per cent.	7·7 per cent.
Body weight	7 st. 2¼ lbs.	7 st. 1¾ lbs.

CASE 42.—A. J.

Diet.	Daily Urine.	Sugar.	Weight.
With potatoes, 11 days	87 ozs.	5·7 p. ct.	5st. 1lb.
Without potatoes, 57 days	84·5 ozs.	6·2 p. ct.	5st. 5½lb.
With potatoes, 15 days	77·4 ozs.	4·4 p. ct.	6st. 1lb.
With potatoes & toast, 9 days	83 ozs.	4·35 p. ct.	6st. 2¼lb.

The following case shews that neither eggs, milk in moderation, nor glycerine increase the amount of sugar.

CASE 43.—A. P., aged twenty-five, on admission weighed 8 st. 5½ lbs. He was passing 244 ozs. of urine of sp. gr. 1040, containing 10 per cent. of sugar. During ten days he was dieted and allowed half a pint of milk, gr. j. extract of opium three times a day, Vichy water and lemon juice.

His urine fell to 174 oz. daily, sp. gr. 1036, containing 5·3 per cent. of sugar, while he gained a stone in weight. During the next ten days he was allowed 1½ oz. of glycerine daily; his urine was 173 ozs. per diem, sp. gr. 1036, with 5·2 per cent. of sugar. These figures are averages for each period. Later on the milk and glycerine were stopped during twenty-four days, and the average quantity of urine in the last week of that period was 174 ozs., sp. gr. 1030, with 6·8 per cent of sugar.

The general effect of strict diet is to cause a diminution of the amount of urine varying from 30 to 50 per cent. in amount, with increase of body-weight and strength, and diminution of thirst. The increase of body-weight is most striking, the patient just quoted gained a stone in ten days, another gained 4½ lbs in fifteen days, and all who improve gain weight.

Some years ago *skim-milk* was brought forward by Dr. A. Scott Donkin as a specific for the treatment of diabetes. He alleged that on six pints of skim-milk daily the sugar disappeared entirely from the urine, and he published cases in support of his statements not only in elderly diabetics but in young persons. Some of his cases proved fatal, but nevertheless where the skim-milk was adhered to the sugar was stated to be greatly diminished or absent altogether. Unfortunately the same success has not attended the practice of other physicians who have tried this plan. Frerichs condemned it, and Roberts thinks one of his patients died from its use. Ralfe speaks of it as only suited to the "gouty type" and in such cases it occasionally does good (Lindsay Porteous). Six pints of skim-milk constitute a very poor diet for people who are suffering from a wasting disease, therefore though I have used a certain amount of skim-milk or pure milk in addition to other diet, I have kept only two patients strictly to this *régime*. One of these unfortunately died. He was four days on this diet, but on the third day the premonitory symptoms of coma (epigastric pain, sighing respiration and rapid pulse) shewed themselves, and proved fatal on the fifth day. This melancholy accident may not be fairly attributable to the skim-milk, but the daily record shews that the diet proved ineffectual to control either the quantity of urine or the amount of sugar.

Date.	Quantity of Urine.	Sp. gr.	Sugar.	Diet.
October 21.	264 ozs.	1035	6·9 per cent.	Ordinary.
,, 22.	212 ozs.	1033	—	—
,, 23.	184 ozs.	—	—	Skim milk.
,, 24.	216 ozs.	1033	6·25 per cent.	—

He grumbled very much, and so did the other patient who fortunately did not meet with any accident. The result in his case is subjoined. While on the skim-milk he lost 3 lbs. in weight, which he regained in a few days on meat diet.

Date.	Quantity of Urine.	Sp. gr.	Sugar.	Diet.
October 16.	184 ozs.	1035	10 per cent.	Ordinary.
,, 17.	130 ozs.	—	—	—
,, 18.	90 ozs.	1038	—	Skim milk.
,, 19.	96 ozs.	—	—	—
,, 20.	104 ozs.	1040	7 per cent.	—

I think that the dietetic value of milk is exaggerated, and that its use as the sole means of nourishment in protracted diseases is unwise. It has been observed that typhoid fever patients keep their strength better and convalesce more rapidly when they are allowed bread and milk than when kept on plain milk. In Bright's disease I have quite given up the use of absolute milk diet, with advantage to my patients, and even in gastric ulcer I push on to a more substantial diet as fast as I can, with the best results; the pendulum has swung too far and too long in the other direction.

Having rejected the skim-milk plan and decided to begin with lean meat, green vegetables, and gluten bread, let us see what other articles we may add to this fare. The following diet-table includes all that is permissible.

Dietary for Diabetics.—Clear meat soup, beef-tea, fish of all kinds, butcher's meat, poultry, game, ham, bacon, smoked meat, eggs, cheese, green vegetables, gluten bread, milk, cream, custard, jelly, blancmange, tea, coffee, claret, Burgundy, brandy,* and whisky.*

No sugar must be used in the preparation of any of these articles.

Those marked with a star (*) must be used in strict moderation.

Fat, in the shape of the fat of meat, butter, or cream, does not cause any appreciable increase in the sugar, and constitutes an important addition to the diet. My experiments with cod-liver oil are quite conclusive that it may be given where there is emaciation with distinct advantage to the patient.

We next come to the important question of what diabetics may drink. In the first place no restriction should be placed upon the quantity. Thirst is the result of a natural craving of the tissues for water, and I would flood the tissues of a diabetic if I knew how to do so.

This primary indication is most simply fulfilled by allowing aërated distilled water, sold as Salutaris, but which can be supplied by any maker of aërated waters. This is a very useful drink in gout, Bright's disease, and other urinary derangements. But as the blood in diabetes is poor in alkali, and it has been suggested with some degree of probability that diabetic coma may be due to the toxic effects of an acid, we may prescribe a bottle of Vichy water (Célestins) or some other alkaline drink daily, with or without lemon juice as preferred, and (aërated) distilled water as much as is desired.

All the alkaline mineral waters, Vals, Vichy, Bourboule, Royat (source Victor), Carlsbad, etc., have a certain repute in the treatment of diabetes, many patients resorting annually to these famous sources for the cure. They would be much wiser to stop at home and order the waters from a druggist. Climate has no influence on diabetes, while travelling, and its inevitable consequences, excitement and fatigue, are the most common causes of its fatal termination.

A good substitute for sugar is the newly-discovered carbon-compound called saccharin $C_6H_4 \langle \begin{smallmatrix} CO \\ SO_2 \end{smallmatrix} \rangle NH$ (Benzoyl sulphonic amide). It passes unchanged through the body, but is intensely sweet. It is very slightly soluble in water, but the addition of a little bicarbonate of soda gets over this difficulty. It is not expensive, as the quantity required is small. Formerly I used glycerine, and my observations confirm those of Küssmaul that it does not increase the secretion of sugar.

In some cases no further treatment is needed to secure the attainment of the four objects I have put before you.

When there is much emaciation cod-liver oil should be given in teaspoonful doses three times a day, and I have already explained and defended this practice.

BATHS.—Diabetic patients often derive great comfort from bathing. A hot-water bath is the simplest, and is most suitable for severe cases. Fat diabetics may use the Turkish bath with benefit. The Russian or vapour bath is also valuable. This treatment is specially indicated where the skin is dry and irritable.

Massage has been strongly recommended by Finkler, as not only causing a general improvement in health and strength, but as leading to a diminution of sugar, which in one case entirely disappeared. Aye has more recently published a case in which considerable benefit followed this mode of treatment, so that he was able to give up his dietetic regimen and discontinue his opium. The plan is worthy of adoption, but must be used with care, and not too much expected from it.

Drugs.—We must now pass to the important subject of the specific treatment of the disease by drugs. Two things are to be expected from their use—the limitation of the urinary secretion, and the diminution of the amount of sugar.

Of many remedies suggested there are four which claim the chief share of our attention. Opium is the oldest, and still retains the first place; next comes its alkaloid, codeine; then salicylic acid or its salts; and lastly, salts of bromine.

The difficulty of carrying out strict comparative experiments with drugs is very great, because over and above the time and labour involved it is oftentimes impossible to secure the necessary conditions for their success, viz., that the other factors of the experiment shall remain undisturbed—an attack of diarrhœa, a severe attack of toothache (to which diabetics are very liable), inability to continue the same diet, and other accidents, interfere with the course of the experiment, and oblige us to reject cases as untrustworthy upon which a good deal of time has been spent.

The following table shews the effect of opium as compared with diet alone and with other drugs. The other drugs used were arsenite of bromine, salicylate of sodium, bromide of potassium, lithia, codeine, salicin, boracic acid, and uranium nitrate.

The case was that of A. P., aged twenty-five, already quoted (Case 43).

Ordinary diet			Diabetic diet.			Minimum without opium.			Minimum with opium.		
Total Urine.	Sp. gr.	Sugar p. c.	Total Urine.	Sp. gr.	Sugar p. c.	Total Urine.	Sp. gr.	Sugar p. c.	Total Urine.	Sp. gr.	Sugar p. c.
244	1040	10	200	1034	8	170	1030	8	176	1030	6·6

This man took two grains of extract of opium three times a day, a much larger dose than usual.

The next case, E. H., aged twenty-seven, shews a more striking result, but the only drug with which opium was compared was salicylate of sodium.

Ordinary Diet.			Diabetic Diet.			Minimum Salicylate of Sodium.			Minimum with Opium.		
Total Urine.	Sp. gr.	Sugar p.c.	Total Urine.	Sp. gr.	Sugar p.c.	Total Urine.	Sp. gr.	Sugar p.c.	Total Urine.	Sp. gr.	Sugar p.c.
186	1042	5·8	158	1031	4·76	110	1038	4·76	58	1034	4·5

The dose was one grain of extract of opium twice daily.

Codeine, when given even in much larger doses than are suggested by its principal advocates, is comparatively inert. Opium is especially valuable for the influence it appears to exert in diminishing the amount of urine. An opium pill often gives an undisturbed night to patients who without it have to rise frequently to pass water. Villemin recommends the combination of belladonna with opium.

The following cases further illustrate the beneficial effects of opium :—

CASE 44.—*Diabetes Mellitus — thirst — polyuria — no gout — disappearance of sugar with diet and opium.*

Mr. William L., aged sixty, consulted me on August 31st, 1886, complaining of *frequent and scanty* micturition. This had been worse for two or three months, but had occurred at intervals for the last two or three years. He never had gout, nor was there any gout in his family. He weighed 14 st. 5 lbs., and had not lost weight within the year. He was obliged to get up very often at night to make water; his total urine was about three quarts in twenty-four hours, amber, faintly acid, sp. gr. 1029, loaded with phosphates and sugar. He was very thirsty, and drank claret and claret and water latterly. Formerly he had been in the habit of drinking beer. Tongue deep red, rather furred. Had formerly suffered from indigestion. On anti-diabetic diet, with Vichy water (half-pint) twice daily with lemon juice, the quantity of his water diminished, his thirst was allayed, and he rested better at night, while during the day he made water less frequently. The urine was, however, still loaded with sugar. He had been getting gluten bread. He was ordered extract of opium (gr. i.) every night at bedtime. A fortnight later he had lost three pounds in weight since his first visit, but on the other hand his urine was 1015, free from sugar, and he could lie in bed all night without being disturbed. He was complaining of costiveness and drowsiness, so the opium was stopped and a euonymin and aloes pill substituted. This gentleman is believed to be alive and fairly well, but has not consulted me since.

CASE 45.—*Diabetes Mellitus—polyuria—sugar fluctuating greatly in amount—hard lump at root of penis—knee jerks present—disappearance of sugar on diet and opium—failure of jambul and antipyrin.*

James A., sixty-one, medal maker, attended as an outpatient on Oct. 29th, 1889, complaining of passing nine or ten pints of water. He weighed 10 st. 9½ lbs. His knee jerks were present. On anti-diabetic diet and extract of opium (gr. i. three times a day) his urine diminished in quantity, but he lost weight, and by the following August weighed only 10 st. 1 lb., although the urine had come down to three pints, and was free from sugar. He began to eat whole-meal bread and other farinaceous food, and rapidly regained his weight, being 10 st. 11 lbs. on Sept. 27th, when he was passing a quart of water free from sugar. After this he continued to gain weight, but the sugar returned.

On Feb. 21st, 1888, he complained of a painless, hard lump at the root of the penis.

In August, 1888, he had a little pain in the right side, and friction was heard. This passed off.

In March, 1889, he weighed 11 st. 4 lbs., was passing about a quart of water daily, sp. gr. 1028, containing a little sugar; the lump at the root of the penis had gone away. He was taking extract of opium (gr. i.) three times a day. He was ordered two of Christy's jambul perles three times a day, and in the course of the next three weeks these were increased gradually to four three times a day, with the result that the urine rose to three pints, and became loaded with sugar. He was then given antipyrin (gr. x.) three times a day with no better result.

CASE 46.—*Diabetes Mellitus—cured by diet and opium.*

Henry H., aged fifty-seven, collier, attended as an out-patient on Jan. 4th, 1887, complaining of thirst and of passing a large quantity of water, seven pints daily. He was put on extract of opium (gr. i.) three times a day, and anti-diabetic diet. By March 22nd the quantity of urine had fallen to two and a half pints, sp. gr. 1015, and the sugar had disappeared. He was allowed to take modified diet, viz., bread and potatoes, and on April 19th the absence of sugar was confirmed.

CASE 47.—*Diabetes Mellitus—polyuria—thirst—wasting —duration three weeks—improvement on diet and opium.*

Emily W., aged twenty-three, married, attended as an out-patient on Dec. 14th, 1886, complaining of great thirst for the past three weeks, of passing a large quantity of water for a month or more, and of wasting. She had been married thirteen months, and six months previously was confined, the baby living only a month. She had become regular again. Her father was dead; he drank too much, and died of liver disease. Her mother died in child-birth. She had only one brother, who was alive and well. Her urine was pale, slightly turbid, acid, sp. gr. 1032, loaded with sugar, and slightly phosphatic. She weighed 7 st. 1 lb. Under treatment by anti-diabetic diet, and extract of opium (gr. i.) three times a day, the quantity of urine diminished, thirst improved, but the sugar remained abundant. On Feb. 1st, 1887, she ceased to attend.

CASE 48.—*Diabetes Mellitus—temporary disappearance of sugar under dietetic restrictions and opium.*

George E., aged sixty-one, attended as an out-patient on Dec. 6th, 1887, complaining of thirst and of getting up a dozen times at night to make water. Quantity of urine five pints, pale, clear, sp. gr. 1027, contained a moderate quantity of sugar. Weight of patient 218 lbs. He had been ill two years. On anti-diabetic diet and extract of opium (gr. i.) three times a day, by May 8th the quantity of water had fallen to three pints, of sp. gr. 1018, free from sugar. He then began to eat bread freely, and the sugar returned, though the quantity of urine remained low. Jambul and antipyrin were tried without effect on the sugar.

CASE 49.—*Diabetes Mellitus—wasting—polyuria—thirst —improvement on diet and opium.*

Benjamin W., aged fifty-three, was admitted as an out-patient on Dec. 13th, 1887, complaining of loss of weight, thirst, and of passing a large quantity of water. He used to weigh 12 st., and now weighed only 9 st. 10 lbs. He passed nine and a half pints of water, sp. gr. 1038, loaded with sugar. On opium and modified diet—for he was not very strict, and would drink a little beer and eat a good deal of brown bread—his urine diminished; on the last occasion he was seen (Nov. 1888) he weighed 9 st. 13 lbs., and was passing only two quarts of water, of sp. gr. 1034, though it was still loaded with sugar.

I have treated many cases with and without opium, and the strong impression left on my mind by these cases, is that in opium we possess the most valuable, and I may add the only trustworthy drug for the treatment of this disease.

Salicylate of sodium as an alkali may be of some service, for I am strongly in favour of the alkaline treatment of diabetes, but I have never observed it to produce any specific effect on the quantity of water or the amount of sugar.

Arsenite of bromine, whether used pure or in the form of Clemens' solution* has never in my hands justified the praises it has received in some quarters. I have already quoted a case (Case 40.) in which it was quite ineffective, if not actually prejudicial, but I have used it in many other cases without being able to observe any specific effect. I have given it in doses of ⅛ gr. of the pure drug, and up to ten minims of the solution. Another form in which arsenic may be given is that known as Martineau's specific.†

Bromide of potassium does not deserve the name of a specific, but is, in my opinion, the best routine remedy to employ in conjunction with the use of opium. It is suitably given in a mixture combined with a little bicarbonate of potassium and some bitter infusion, and very satisfactorily allays the nervous irritability so often present.

---

* *Liquor Clementis.*—According to Hager its composition is as follows: *Acidi Arseniosi, Potassii Carbonatis,* āā gr. jss. Place in a test tube and add five drops of distilled water, and warm so as to form a clear solution. Dilute this with distilled water till the quantity weighs 158 grains. Then add 6·2 grains (4 drops) of bromine and set aside for one day. The fluid is then ready for dispensing. A modified formula is frequently employed now.

† *Carbonate of Lithium,* 20 centigrammes (3 grains), added to a tablespoonful of the following solution:—*Arseniate of Sodium,* 0·20 centigrammes ($\frac{1}{30}$ grain); *Distilled Water,* 500 centigrammes (80 minims); the mixture to be placed in an ordinary soda-water machine (Briet's apparatus), and this quantity to be used daily with and between meals, as a drink, alone or mixed with wine.

Teissier has found it very useful in the treatment of the diabetes of elderly women in doses of thirty to forty-five grains daily; but we know that this is a mild form of diabetes, with a natural tendency to amelioration or cure.

Among other remedies recommended are *antipyrin* in 15 grain doses thrice daily (Germain Sée); *sulphide of calcium*, in doses of ¼ to ⅓ grain, three or four times in twenty-four hours (Cauldwell, Draper); *creasote*, 4 to 10 drops daily (Valentine); *phenacetin* and *exalgine* (Dujardin-Beaumetz); *camphor* (Peyraud), *iodoform* (Moleschott); *nitro-glycerine* (Kennedy); *salicin*, in doses of from 2 to 18 grammes (30 to 270 grains) daily (Dornblüth); *carbonate of soda*, ½ oz. to 2 ozs. daily, with *citric acid* (Stadelmann); 5 to 6 ozs. daily of a 10 per cent. infusion of the bark of *syzygium jambolanum* (Quanjer); *sulpho-carbolate of soda*, 5 to 30 grains for a dose (Monckton); *pilocarpine* ($\frac{1}{20}$ grain), with 1 grain of sulphate of potash, three times daily (Eager); *nitric* or *nitro-hydrochloric acid*, with tincture of *nux vomica* (Wilks); *cocaine* (¼ grain), three times a day (Oliver); *pepsin* (10 grains), three times a day (Gardner); *jambul* (powdered bark of *eugenia jambolana*), in doses of 3 to 5 grains, thrice daily (Kingsbury); *ouabaïn*, in doses of $\frac{1}{1000}$ of a grain three or four times a day (Gemmell); *phosphorus*, in perles containing each $\frac{1}{30}$ of a grain, of which three and then six were taken daily (Balmanno Squire).

Jambul.—This drug was originally recommended by Banatvala of Madras, and spoken of in terms of highest praise—more recently Quanjer in Java has had an equally favourable experience of it. Both these writers mention the syzygium jambolanum as the source. Kingsbury, who drew the attention of the profession in this country to its use by the recital of some successes he had met with, speaks of the plant as eugenia jambolana.

On consulting Messrs. Christy, I am informed that these plants are the same.

In these observations the drug was administered in the form of Christy's jambul perles.

CASE 50.—A. J., aged eighteen:—

Diet.—Meat, greens, Bonthron's gluten bread, and one pint of milk. This was not changed throughout.

Dose.	Duration.	Urine.	Sugar.	Body Weight.
Morphine—				
¼ grain thrice daily	7 days	85 oz.	9 per cent.	5 st. 4¾ lbs.
Jambul—				
6 perles daily	3 days	100 oz.	7·6 per cent.	
9 perles daily	6 days	100 oz.	6·8 per cent.	
12 perles daily	6 days	104 oz.	8 per cent.	5 st. 6½ lbs.
Ext. Opii—				
3 grains daily	7 days	76 oz.	5·4 per cent.	5 st. 10 lbs.

CASE 51.—W., aged thirty-three:—

Diet.—Meat, greens, Bonthron's gluten bread, potatoes, one pint of milk.

Dose.	Duration.	Urine.	Sugar.	Body Weight.
Morphine—				
¼ grain daily	6 days	167 oz.	9 per cent.	8 st. 5½ lbs.
Jambul—				
6 perles daily	6 days	205 oz.	5·2 per cent.	8 st. 5¼ lbs.
9 perles daily	5 days	285 oz.	5·8 per cent.	
12 perles daily	4 days	394 oz.	7·6 per cent.	8 st. 3 lbs.
Ext. Opii—				
3 grains daily	7 days	200 oz.	6·6 per cent.	8 st. 8½ lbs.

In another case as many as thirty-six perles were given daily without effect, and a liquid extract also supplied by Messrs. Christy was given in several cases in doses gradually rising to 1½ ozs. daily.

It was tried in every other case under treatment about the same time, and in no single instance was any distinct benefit observed to follow its use.

During the administration of *ouabaïn* for whooping-cough, sugar was observed to disappear from the urine of one of the cases on which it was tried (Gemmell).

Messrs. Christy kindly supplied me with a sample of *liq. ouabaïni* prepared by them, which I administered to several diabetics without observing any benefit. The dose was carried as high as nine drachms daily, beyond which I did not venture to go.

Ouabaïn is an alkaloid having the following composition:—$C_{30}H_{46}O_{12}$; it is obtained from the ouabaïs plant which is used as an arrow poison by the Somalis on the East Coast of Africa.

Phosphorus, which was recommended by Balmanno Squire because he had observed the sugar disappear during its use from the urine of an elderly man suffering from eczema, has had an extensive trial without any evidence being obtained that it possesses the power to control any of the important symptoms of diabetes, and exactly the same may be said of cocaine, which was recommended by Oliver on theoretical grounds.

Defresne, basing his treatment on the pancreatic theory of diabetes, recommended *pancreatine* in very large doses (8 grammes daily), and Lépine has employed *pilocarpine*, given subcutaneously to stimulate the pancreatic secretion.

TREATMENT OF SYMPTOMS.—*Neuralgia*.—In many instances, as in cases quoted, diabetic neuralgia ceases when the disease is checked by general treatment. But to afford relief we may give antipyrin, 15 or 20 grains for a dose, or exalgine, 3 to 4 grains for a dose, or opium or morphia, by the mouth or hypodermically. The following ointment may be used locally: ℞ *Menthol*, gr. xv.; *Cocaine*, gr. v.; *Chloral Hydratis*, gr. iij.; *Vaseline*, 3j.; *Fiat Unguentum*. Sig.—To be rubbed gently over the seat of pain.

Thirst depends on the great drain of water from the body, and the rational treatment of it is to lessen the polyuria. In the meantime no good purpose is served by placing restrictions on the quantity of fluid to be drunk, as this can only augment the patient's suffering without any benefit. Sucking ice, or the use of lemon or bitartrate of potash water as a drink, is useful to check thirst. With the object of promoting a flow of saliva and so relieving the dryness of the mouth, hypodermic injections of pilocarpine (grain $\frac{1}{20}$ to $\frac{1}{6}$) have been frequently used, and with good result; McAvoy recommends for the same purpose 5 minim doses of glycerinum acidi carbolici. Constipation appears in some cases to increase thirst.

A subordinate but nevertheless highly important detail of treatment is the regulation of the bowels. All, or almost all, diabetics are *constipated*, whatever they may say. I have seldom made a post-mortem examination of a subject of this disease without finding the colon stuffed with hardened masses of fæces. Therefore, it is usually necessary to give a purgative twice a week. We may use any of the favourite or fashionable remedies, avoiding only such preparations as contain sugar.

The *œdema* of diabetes generally disappears very rapidly on rest in bed and appropriate general treatment. Dickinson thinks large doses of perchloride of iron (30 to 40 minims of the tincture three or four times a day) act as a specific. Massage should be tried in all obstinate cases.

Pruritus.—This may be treated by careful washing with some antiseptic fluid. Lawson Tait speaks very favourably of the use of a lotion containing hyposulphite of soda, an ounce to a quart of water, and of an ointment containing 10 grains of sulphuret of potassium to the ounce, combined with the use of opium. A warm, freshly-made saturated solution of boracic acid may be used with confidence that it will give relief, and after carefully drying the parts, zinc ointment should be applied freely, but it is right to say that the condition, possibly from want of care on the part of the patient, is sometimes very intractable.

Routh recommends the following remedies: Put a teaspoonful of borax into a pint bottle of hot water, add 5 drops of oil of peppermint and shake well. Bathe the affected parts freely with a soft sponge. If rawness or eczema is present, apply olive oil to which iodoform has been added in the proportion of 5 grains to the ounce.

Sweating.—This is a rare affection, and according to Schmitz, is a favourable symptom, so that it may not be always desirable to check it. But where this has ceased to be the case I have found that Dover's powder, taken as a substitute for the opium preparation previously in use, proved very useful.

BIBLIOGRAPHY.

ARMITAGE (W. S.). Semolina in Diabetes. "Lancet," 1883, Vol. I., p. 859.

AYE. Gymnastic Treatment of Diabetes. "Berlin. Klin. Woch.," 1889, No. 30.

BANATVALA. The Use of the Syzygium Jambolanum in Diabetes. "Mid. Med. Miscell.," 1883.

CAULDWELL (C. M.). Calcium Sulphide in the Treatment of Diabetes Mellitus. "New York Med. Jour.," 1884, Vol. I.

DEFRESNE (M. Th.). Essais sur le Mécanisme du diabète maigre. "Gaz. des Hôpitaux," 1890, No. 59.

DONKIN (A. S.). The Skim-milk Treatment of Diabetes and Bright's Disease. London, 1871.

DORNBLÜTH (O., Junr.). Ein Beitrag zur Theorie und Praxis der Azneibehandlung des Diabetes mellitus. "D. Archiv. f. klin. Med.," Bd. XXXVII., Heft 1 and 2.

DRAPER. Sulphide of Calcium in Diabetes. "New York Med. Jour.," 1884, Vol. I., p. 380.

DUJARDIN-BEAUMETZ. Du Régime alimentaire dans le Diabète. "Bull. de Thérap.," Vol. III., p. 385.

EAGER (T. C.). A case of Diabetes treated by Pilocarpine. "Lancet," 1884, Vol. II., p. 275.

FINKLER. The Treatment of Diabetes by Massage. "Fifth Congress for Internal Med.", Wiesbaden, 1886.

GARDNER (E. B.). Pepsin in Diabetes. "Practitioner," 1886, Vol. XXXVII., p. 448.

GEMMELL (W.). On Ouabaïn in Whooping Cough. "Brit. Med. Jour.," 1890, Vol. I., p. 950.

HAGER. Composition of Liquor Clementis. "Wiener Med. Blatt.," 1883.

KENNEDY (R. A.). Nitroglycerine in Diabetes. "Canada Med. Record," Jan., 1888.

KINGSBURY (G. C.). Jambul Seeds in Diabetes. "Brit. Med. Jour.," 1887, Vol. I., p. 617.

McAVOY (J.). Glycerinum Acidi Carbolici in Diabetes Mellitus. "Lancet," 1885, Vol. I., p. 786.

MOLESCHOTT (J.). Iodoform gegen Diabetes mellitus. "Wien. Med. Woch.," Bd. XXXII., 1882.

MONCKTON (F. A.). Sulphocarbolate of Soda in Diabetes Mellitus. "Australasian Med. Gaz.," 1884.

NÉDATS (De.). Quoted by Dujardin-Beaumetz, *op. cit.*

OLIVER (T.). Cocaine in Diabetes. "Lancet," 1889, Vol. II., p. 785.

PORTEOUS (J. L.). Diabetes Mellitus: Skim-Milk Treatment. "Edin. Med. Jour.," Vol. XXX., Part I., p. 508.

QUANJER. Use of the Bark of Syzygium Jambolanum in Diabetes. "Woekbl. v. h. Nederl. Tijsche voor Geneesk," 1888, Vol. I., p. 251.

SÉE (Germain). Antipyrin in Diabetes Mellitus. "Brit. Med. Jour.," 1889, Vol. I., p. 1472.

SQUIRE (Balmanno). A case of Diabetes benefited by Phosphorus. "Brit. Med. Jour.," 1889, Vol. II., p. 1216; 1890, Vol. I., p. 293.

TEISSIER. Sur le traitement du diabète sucré. "Bull. de Thérap.," Vol. CVIII., p. 332.

VALENTINE (P.). Le créosote contre le Diabète. "L'Osservatore," Jan. 5, 1889.

VILLEMIN. Traitement du Diabète aigu. "Bull. de l'Acad. des Sciences." "Arch. Gén. de Méd.," 1887, Vol. I., p. 364.

WILKS (S.). Cases of Diabetes treated with Nux Vomica and Mineral Acids. "Med. Times and Gaz.," 1884, Vol. I., p. 320.

Part II.—DIABETES INSIPIDUS.

CHAPTER VII.

IN spite of the discovery of Willis, in 1670, of the presence of sugar in the urine in certain cases of diabetes, it was not till the end of the eighteenth century that the formal distinction into diabetes mellitus (verus) and diabetes insipidus (spurius) was made by Cullen.

Diabetes insipidus, according to Senator, is the name applied to every case of chronic morbidly increased secretion of urine, free from sugar, which is caused by no profound structural alteration of the kidneys, and which constitutes either the sole or at least the most prominent and primary morbid phenomenon. According to this definition it is the polyuria which causes the thirst, but he admits that there are cases in which the thirst is primary, and it is preferable to define the disease simply as characterised by the two symptoms, thirst, and polyuria, with absence of glycosuria. The only objection to this is that it may include certain rare cases of cystic degeneration of the kidneys, as such cases have been met with where no albuminuria was observed (Strange). We may therefore add that it is not dependent on structural alterations in the kidneys, though where enlargement of these organs cannot be detected, and no albumen is found, there is great probability that these conditions will be confounded.

ETIOLOGY.

Bernard produced simple polyuria by puncture of the floor of the fourth ventricle near the spot injury of which is followed by glycosuria. Eckhard produced similar results in rabbits by injuring the posterior lobe of the vermiform process of the cerebellum. Peyrani induced the condition by electrical stimulation of the cervical sympathetic; and it is said to follow section of the splanchnic nerves.

Predisposing Causes.—The *age* at which the disease occurs is illustrated in the following table:—

Age.	Roberts.	Strauss.	Van der Heijden.
Under 5 yrs.	7	9	2
5—10	15	12	5
10—20	18		19
20—30	16	57	23
30—40	—		19
40—50	15		9
50—60	—	7	6
60—70	4	—	4
	70	85	87

This shews that the greatest frequency is during adolescence, early manhood and middle life, infancy and old age being comparatively exempt. The male sex seems decidedly more liable than the female in the proportion of nearly three to one:—

Name.	Males.	Females.
Roberts	55	22
Van der Heijden...	71	25
	126	47

The influence of *heredity* is well known. Weil has published a very remarkable history of a family in which no less than twenty individuals out of a total of ninety-one, in four generations, were attacked by the disease. Males and females were attacked in equal proportions, and no age appeared exempt. There was no evidence of scrofula or tubercle in the family; all the males were well developed and had performed their military service. It seemed to have no influence on their vitality, the diabetic ancestor having lived to be eighty-three, his daughter to seventy-four, and two living members, also affected, having attained the ages of sixty-seven and seventy-six respectively.

In Lacombe's case, a mother, her four children, and her brother and his children were all affected.

It has been supposed that the *tubercular* diathesis is a predisposing cause, and although in Weil's family, just quoted, there was no evidence of such a taint, R. H. Clay has recorded the cases of three children, two brothers and a sister, suffering from diabetes insipidus, in whose family four other children had died of consumption.

Syphilis is undoubtedly a cause, but in this case the effect is no doubt due to the formation of a gumma in the brain.

Gout is probably to be reckoned as a predisposing factor, and so must *insanity*.

Among the *exciting causes*, severe *contusions* in various parts, but particularly the head, occupy a well-defined place. One of Matthews Duncan's cases was apparently caused by a blow of great severity on the back of the head. Flatten has reported a case in which a blow from the trunk of a tree on the left side of the neck and occipital region, was followed in seven days by great thirst and polyuria, accompanied by complete paralysis of the left external rectus oculi muscle, and a slight paresis of the corresponding muscle on the right side. He thinks the blow caused a circumscribed hæmorrhagic softening on the level of the point of exit of the abducens nerves in the hindermost part of the pons or the anterior part of the medulla. Drummond, of Newcastle, has described a case in which the disease came on suddenly in a man who had been roughly throttled by a fellow workman, and Oliver mentioned the case of a man who fell from a ship and was rendered unconscious for ten or twelve days; on recovering consciousness he drank greedily, and from that moment polydipsia and polyuria existed.

Nothnägel recorded a case of a man aged thirty-four, who was kicked in the left side of the abdomen and fell with the back of his head on the hard floor. He suffered from pain with a sense of compression in the occiput, and his pupils were much contracted. Soon after he was seized with thirst and passed a large quantity of water. These symptoms gradually disappeared, and eighteen days after his accident he was discharged cured.

The influence of *pregnancy* is also not to be doubted, but the condition appears to be temporary. Thus in Hughes Bennett's case the disease came on at the fifth month and disappeared two days after delivery; in Matthews Duncan's case it disappeared a month after delivery, and in Esterle's the disease appeared in two successive pregnancies, but disappeared in the interval. Westphal has published a case which he attributed to the loss of blood at the confinement. Senator has recorded a case which after lasting several years *disappeared* on the supervention of pregnancy.

Possibly in close alliance with the last-named cause, in their mode of origin, are those cases in which *abdominal tumours* have been associated with this disease. Ralfe has met with it in two cases of *aortic aneurism*, and Haughton in a case of *uterine fibroid*, with enlargement of the mesenteric glands and fæcal accumulation. Ralfe believes the aneurisms pressed upon the vagus.

Somewhat allied, too, is the case described by Küster, which followed the incision and drainage of a *pancreatic cyst*.

Among other exciting causes are exposure to *cold* and drinking cold fluids when heated, previous *febrile* or *inflammatory diseases*, *muscular effort*, exposure to a *hot sun*, and *mental emotion*.

As illustrations of the effect of *muscular effort* may be quoted Frank's case of a boy who strained himself by pushing the wheel of a cart which was sunk in the mud; and Jarrold's of a girl who slipped, but saved herself from falling by a very great exertion.

Jacobi has recorded a case which was cured by the removal of a tænia medio-canellata; but in a case of Lunin's, removal of the worm was not followed by any benefit to the specific symptoms.

Acute alcoholic intoxication is stated by Dickinson to be an indubitable exciting cause; in his own words " a person gets what is vulgarly called dead drunk "; on regaining consciousness he is polyuric and so remains.

Finally, lesions of various parts of the nervous system are direct exciting causes of the disease, but these are better described under the heading of morbid anatomy.

MORBID ANATOMY.

Nervous System.—Various lesions of the *brain* have been observed, such as fracture of the base of the skull, with contusion of the anterior lobes; yellow softening, inflammation and degeneration in the floor of the fourth ventricle. Tumours, tubercle, glio-sarcoma, gummata, and exostoses in various positions; carcinoma of the pineal gland. Lesions have also been found in the semilunar ganglia and splanchnic nerves. The ganglia shewed the atrophy of nerve cells associated with increase of connective tissue and dilatation of vessels, which are seen in so many diseases. The nerves shewed fatty degeneration of their axis cylinders, with granular degeneration and dilatation of the vessels of the neurilemma (Schapiro).

The Blood.—According to Strauss, the solids of the blood are increased:—

	First Observation.	Second Observation.
Water	77·79 per cent.	77·937 per cent.
Solids	22·21 per cent.	22·063 per cent.
	100·00	100·000

The solids in the second case consisted of :—

Fibrin	0·467 per cent.
Hæmoglobin	11·72 per cent.
Other Albuminoids	7·441 per cent.
Extractives	1·301 per cent.
Ash	1·134 per cent.

The serum contained :—

Water	88·712 per cent.
Albumen	9·062 per cent.
Extractives	1·012 per cent.
Ash	1·214 per cent.

Alimentary System.—The mucous membrane of the large intestine may be ulcerated and ecchymosed (Schapiro), or there may be tubercular ulceration (Roberts.) Dickinson has observed cancer of the liver in one instance.

Respiratory System.—Where, as has several times happened, the disease is connected with tubercle, there may be more or less extensive pulmonary phthisis. Lobular pneumonia was present in one case (Neuffer). As a rule the lungs are normal.

Urinary System.—The kidneys appear normal or may be congested, the malpighian bodies especially appearing as well marked hyperæmic points.

Secondary changes in the kidneys are not uncommon. Thus Neuffer describes a case in which the renal tubules were dilated, some being stripped of their epithelium, others stuffed with fatty *débris;* Beale observed similar changes, together with increase of the peri-vascular connective tissue. In some cases the kidney is sacculated, the cavities containing urine or pus, and the bladder and ureters are also thickened and dilated.

PATHOLOGY.—There can be little doubt that the disease is a neurosis which depends upon changes in the nerve centres, or upon peripheral irritation acting through those centres. The channel for the transmission of the morbid influence appears to be the vagus nerve, and the

DIABETES INSIPIDUS. 211

mechanism consists in relaxing the tonus of the renal arterioles and permitting a very free afflux of blood to the malpighian bodies.

SYMPTOMS, AND CLINICAL HISTORY.

As has been already stated, the cardinal symptoms are thirst and polyuria. In certain cases the thirst is the first symptom noticed, in others, and perhaps in the majority, the polyuria is the primary affection.

The following case illustrates very well the type of case in which thirst precedes the polyuria:—

CASE 52.—*Diabetes Insipidus—thirst—polyuria—gouty history—improvement.*

John P., forty-nine, came as an out-patient on April 10th, 1883, complaining of thirst and passing a large quantity of water. He said that his illness began quite suddenly six years before. He was out for a walk one Sunday when "a great thirst came on him, so that at every house or brook he came to he must be drinking." He was in perfectly good health, and had not met with any accident. He had worked as a stamper in Birmingham, and had always enjoyed good health. The only illness he could recall was "stoppage of the water" eighteen years before, but he soon recovered. He was a married man, but had no children living. Two were born dead and his wife had had several miscarriages, but he could give no history of syphilis, sore throat, or skin eruption, though he admitted having had gonorrhœa when he was nineteen. He had no knowledge of any similar case in the family. He had had gout several times. He was a well-nourished man; there was a depressed scar over the right frontal eminence, which was caused by a fall in early childhood before he could remember. He weighed 12 st. 7 lbs. His tongue was furred posteriorly, and he suffered from pain and wind, after food. His bowels were regular; liver and spleen normal. He complained of severe cough and spit, and shortness of breath, but he had no palpitation or œdema of feet; his heart and lungs appeared normal. Pulse 120. He said that he passed a chamber-pot full of water in the night and more in the day. The urine was clear, pale yellow, sp. gr. 1001, faintly acid, no albumen or sugar. Ophthalmoscopic signs negative. He had no headache, giddiness, or vomiting. He was treated by colchicum, carbonate of magnesia, and sulphate of magnesia, and under these remedies his polyuria and thirst greatly diminished.

As will have been noted, this disease does not involve weakness or loss of flesh, but is consistent with good general health. In some cases, there is wasting and evident ill-health, in others obesity (Voss).

When it comes on early in life it affects the growth of the individual. The appetite is sometimes very large (Trousseau), the flow of saliva may be greatly increased (Kuelz), and there are often pains in the back and loins. The bowels are generally confined. The temperature of the body is normal, in some cases it is sub-normal. The following is a good example of this type of case:—

CASE 53.—*Diabetes Insipidus—neurotic family history—stunted growth—polydipsia.*

Frederick S., sixteen, jeweller, was admitted into the General Hospital on June 19th, 1890, complaining of thirst and of passing large quantities of water.

Family History.— Father died aged thirty, from "diseased bowels." Mother alive, aged forty-two, had been an inmate of a lunatic asylum for three years. She suffered from fits before becoming insane. Two brothers living—one died in infancy. No history of phthisis or diabetes.

Previous History.—Patient was always weak. He had measles at six years of age. Ever since his infancy he had passed water in large quantities, and had suffered from thirst. For the last three years he had been in the habit of drinking quite a quart of water during each night. His appetite had never been excessive.

Present Condition.—Patient was fairly well nourished, very small for his age; weight, 4 st. 6 lbs.; height, 4 ft. 7¾ in. No jaundice, œdema, or partial dryness of skin. He seemed very lively and intelligent. Pulse, 80; Resp., 16; Temp., 98·5.

Alimentary System.—There was a red line along the margin of the gums. Tongue moist and furred. Appetite poor. Thirst marked; he drank 6½ pints of fluid yesterday, and one pint in the night. No pain after food, or vomiting. Bowels rather confined. Liver dullness in V.M.L., 3 in. Splenic dullness in M.A.L., 1½ in.

Circulatory System.—There were some dyspnœa and palpitation on exertion; cardiac impulse obscure, in 5th I.S., in V.M.L. Cardiac dulness was bounded above by the 3rd C.C. to the right by the right edge of the sternum, and to the left by the V.M.L. Heart sounds rather feeble; pulmonary second sound reduplicated; no murmur. Pulse regular, fairly strong, easily compressed.

Respiratory System.—No cough. There were some dilated veins over the upper part of the front of the thorax. Sternum rather convex. Vocal fremitus feeble; percussion note resonant; breath sounds vesicular; vocal resonance normal; no adventitious sounds.

Urinary System.—Micturition frequent; usually about three times in the night. No scalding or pain. Urine 104 ozs., sp. gr. 1002, neutral, pale opalescent straw colour; white deposit consisting of bacteria, mucus, leucocytes and triple phosphates; urea 0·65 per cent.; a haze of albumen; no sugar or blood.

Analysis of another specimen shewed total solids to be only 0·66 per cent. These were made up as follows:—

Organic substances—
  Urea   -   -   -   0·324
  Albumen and mucus   0·040 = 0·364 per cent.
Inorganic substances—
  Chlorides   -   -   0·023
  Sulphates   -   -   0·044
  Soluble Phosphates   0·060
  Insoluble Phosphates   0·020
  Magnesia, Lime, Soda,
    and Potash   -   0·149 = ·296 per cent.

                  Total   0·660 per cent.

The following table shews the quantity of fluid ingested and the quantity of urine passed in corresponding periods of twenty-four hours, during his stay in hospital.

Date.	Urine in Ounces.	Fluid ingested in Ounces.
June 22	104	130
,, 23	110	120
,, 24	126	100
,, 25	106	100
,, 26	116	120
,, 27	108	120
,, 28	122	140
,, 29	118	130
,, 30	136	150
July 1	160	150
,, 2	140	140
,, 3	120	130
,, 4	128	110
,, 5	130	110
,, 6	120	130
,, 7	122	110
,, 8	128	120
,, 9	100	110

There was a disposition for his temperature to be subnormal, ranging 97·5 to 98°. He gained no weight while under treatment, nor was the urine affected by the drugs employed, viz., a mixture of dilute hydrochloric acid with a bitter infusion, given well diluted with water as a drink, and 3 grains of ergotin daily, in the form of pills, containing each 1 grain.

These patients suffer a good deal from dryness of the skin, and dry mouth and fauces. According to Lowinsky boils, originating in inflammation of the sweat glands, may occur. The bladder may become very irritable, and œdema of the feet may supervene in the later stages.

Great intolerance of alcohol has been observed, which is the more remarkable as it must be excreted very rapidly, while Trousseau has recorded a case in which great tolerance existed, the patient being able to drink twenty litres of wine and a litre of brandy at a sitting.

Hunger may be present in a very marked degree, and Trousseau speaks of these patients as the terror of the keepers of restaurants *à prix fixe*.

The *Urine* is always greatly increased; it may be faintly acid, alkaline, or neutral; it is always of low specific gravity, and as a rule free from albumen. It contains no glucose, though cases in which diabetes insipidus appears to alternate with diabetes mellitus have been observed. Inosite, or muscle sugar, has been frequently found to be present (Schulzen, Strauss). It is detected by evaporating the urine to dryness, then moistening with ammonia and a solution of calcium chloride, and evaporating again. If inosite is present a ruby red colour makes its appearance (Scherer).

The polyuria is occasionally intermittent (Fagge).

The urea is generally increased; the uric acid is probably normal; kreatin and kreatinin are said to be diminished (Strauss, Pribram); hippuric acid is present (Bouchardat); the chlorides are increased (Senator); and so are the phosphates.

Intercurrent febrile diseases usually produce a diminution in the quantity of urine. This has been observed in small-pox (Lacombe, Charcot, Kuelz), typhus (Pribram), pleurisy (Desgranges), acute rheumatism (Roberts), pneumonia and erysipelas (Senator). But Dickinson has recorded a case in which an attack of scarlatina was attended by no diminution.

Affections of the *eye* do not form any part of the proper symptoms of this disease. Cataract is rare, and such conditions as double optic neuritis, atrophy of the optic nerve (Gowers), retinal hæmorrhages (Galezowski), and staphyloma posticum (Laycock), are either the consequences of some cerebral lesion, or accidental coincidences.

PROGNOSIS.—The *duration* of the disease is very indefinite, as it may exist the whole of a long lifetime, and does not appear to have any essential tendency to cause death.

Many instances of spontaneous cure have been recorded; thus a case that had existed eighteen years was cured by an attack of acute rheumatism, another by a pleurisy treated by blistering, and one already alluded to, by the removal of a tapeworm. Garnerus has recorded the case of a child who soon after birth suffered from polyuria with progressive emaciation, and during the second month sugar appeared in its urine. This disappeared later on, but the polyuria did not improve till the child was fed on milk from which the sugar was removed by fermentation, sweetened with glycerine and mannite and diluted with boiled water. After three months of this diet ordinary milk was given it without any relapse following, and the child remained well.

The prognosis as to life is not unfavourable, as has been already shewn in Weil's cases. The cause must determine this to some extent; obviously if there is reason to believe in disease of the central nervous system, the prognosis must depend on the curability of this condition. Where there is a history of syphilis a cure may be hoped for. We have seen that after injuries the condition may be temporary, and in pregnancy it terminates as a rule, at or soon after delivery. While the disease may exist for years without apparent alteration of health, when the kidney changes, described in the section on morbid anatomy, set in, the patient eventually dies with symptoms of chronic uræmic poisoning, therefore evidences of renal affection, such as albuminuria, must be regarded as unfavourable.

TREATMENT.—In the majority of chronic idiopathic cases drugs are of little use. Where there is a syphilitic history mercury and iodide of potassium should have a prolonged trial.

One of the most highly-recommended remedies is valerian, given either in the form of extract, in drachm doses, or valerianate of zinc, up to 15 grains, three times daily.

Ergot, in the form of the liquid extract, 10 to 30 minims, three times a day, or ergotin, in doses of 1, 2 or 3 grains, three times a day, have also been successful.

Eichhorst has cured a case with tincture of the acetate of iron, and another with 5 grammes (75 grains) of antipyrin, daily. The latter drug has been also successful in the hands of Laplane, Zenner, and Opitz.

Galvanisation applied to the nape of the neck (Althaus, Clubbe) has been strongly recommended.

Kennedy has urged the use of nitric or nitro-muriatic acid in doses of from 1 to 5 drachms daily. Libby has employed with advantage spirits of turpentine, 15 to 20 drops, three times a day. Murrell observed benefit in a traumatic case from a combination of ergot with belladonna. Roberts recommends a blister to the pit of the stomach.

It is useless and unnecessarily cruel to attempt to limit the quantity of fluid drunk. The dryness of the mouth may be relieved by pilocarpin, or by sucking ice, or lemon, or glycerine lozenges. Opium or morphia may be tried to diminish the polyuria, but obviously the general condition of the patient must be our guide as to the extent to which we should push active treatment. While the general hygienic surroundings of the patient should be as satisfactory as they can be made, there is no reason to make him more of an invalid than he need be, and his diet, while wholesome and sufficient, should not be restricted.

The dryness of the skin may be relieved by frequent steam or hot water baths, and there is some advantage in anointing the skin after the bath with olive oil, scented by the addition of a few drops of rosemary or bergamot.

BIBLIOGRAPHY.

ALTHAUS (J.). Diabetes Insipidus treated by Galvanisation of the Medulla. "Med. Times and Gaz.," 1880, Vol. II., p. 617.

BENNETT (J. HUGHES). The Principles and Practice of Medicine. 5th Edition. Edinb., 1868, p. 996.

CLAY (R. H.). Three cases of Diabetes Insipidus in one family. "Lancet," 1889, Vol. I., p. 1188.

CLUBBE (C. P. B.). Diabetes Insipidus treated by Electricity. "Lancet," 1881, Vol. II., p. 749.

DRUMMOND. Proceedings of the Northumberland and Durham Medical Society. "Brit. Med. Jour.," 1889, Vol. II., p. 928.

DUNCAN (J. MATTHEWS). On Diabetes Insipidus in Pregnancy and Labour. "Obstet. Trans.," Vol. XXIV., p. 308.

EICHHORST. Praktische Erfahrungen über die zuckerige und einfache Harnruhr. "Corresp. Blatt. fur Schweizer Aerzte," 1888, No. 13.

ESTERLE. Quoted by Tarnier and Budin. "Traité de l'art des Accouchements." Paris, 1886, p. 51.

FAGGE (C. H.). The Principles and Practice of Medicine. Vol. II., p. 347.

FLATTEN. Beitrag zur Pathogenese des Diabetes insipidus. "Arch. f. Psych.," 1882, Bd. XIII.

GARNERUS. Combined Diabetes Mellitus and Insipidus in a new-born child : cure. "Deut. med. Woch.," Oct. 23, 1884.

HAUGHTON. Uterine Tumour with Enlarged Mesenteric Glands. "Dub. Quar. Jour.," 1863, p. 323.

JACOBI. Diabetes Insipidus. "Arch. of Pediatrics," Nov., 1888.

KENNEDY (H.). Cases of Diabetes Mellitus, and the Treatment found Successful. "Pract.," 1878, Vol. XX., p. 94.

KUELZ. Studien über Diabetes mellitus und insipidus. "D. Archiv. f. klin. Med.," Bd. XII., p. 248.

KÜSTER. Zur Diagnose und Therapie der Pancreascysten. "Deut. med. Woch.," 1887, Bd. XIII.

LAYCOCK (T.) Beneficial use of Jaborandi in cases of Diabetes Insipidus or Polydipsia. "Lancet," 1875, Vol. II., p. 242.

LOWNISKY (J.). Furunculose bei Diabetes insipidus. "Cent. f. klin. Med.," 1890.

LUNIN. Ueber Diabetes Insipidus. "Jahr. für Kinderheil.," Bd. XXI., Heft 4.

MURRELL (W.). On a case of Diabetes Insipidus treated by (1,) Belladonna ; and (2,) Ergot. "Brit. Med. Jour.," 1876, Vol. I., p. 8.

NEUFFER. Ueber Diabetes insipidus. "Diss. Tübingen," 1856.

NOTHNÄGEL. Durst und Polydipsie. "Virchow's Archiv.," 1881, Bd. LXXXVI.

OPITZ (M.). Antipyrin in Diabetes insipidus. "Deut. Med. Woch.," Aug. 8, 1889.

RALFE (C. H.). Two cases of Aortic Aneurism, with Increased Secretion of Urine. "Lancet," 1876, Vol. I., p. 308.

SCHAPIRO. Zur Lehre von der zuckerlösen Harnruhr. "Zeitschr. f. klin. Med.," Bd. VIII., p. 191.

STRAUSS. Die Einfache zuckerlöse Harnruhr. "Prager Vierteljahrsch," 1871, Bd. CXIII., p. 1.

TROUSSEAU (C.). Clinique Médicale de l'Hôtel-Dieu de Paris. Quatrième Edition, p. 808.

VOSS (A.). Uber Diabetes insipidus und adipositus. "Berlin. klin. Woch., 1891, No. 1.

WEIL (A.). Ueber die hereditäre Form des Diabetes insipidus. "Virchow's Archiv.," Bd. XCV., p. 70.

WESTPHAL (A.). Ein Fall von Diabetes insipidus. "Berlin. klin. Woch.," 1889.

ZENNER (P.). A case of Diabetes Insipidus with favourable termination. "Cleveland Med. Gaz.," Sept., 1889.

GENERAL INDEX.

ABRAHAM, p. 56; Albert, 86; Albertoni, 73; Addison, 99; Almèn, 13; Althaus, 59, 107, 217; Anderson (Wallace), 110; Andral, 104; Apt, 66; Armanni, 71, 72; Armitage, 186; Aretæus, 2; Arthaud, 10, 12, 15, 58, 59, 60; Ashdown, 5, 13; Auerbach, 30, 107; Aye, 192.

	PAGE
Abdomen, Malignant Disease of	36
Abdominal Tumours	208
Abscess of Liver	65
— of Pancreas	67
Acetate of Iron	217
Acetone in Urine	88, 136
— Tests for	91, 142
Aceto-acetic Acid in Urine	88
Acetonæmia	63, 139, 142
Acetonuria	142
Acids, Researches into the effects of	144
Acne Pustules	94
Acute Febrile Glycosuria	80
— Rheumatism, as cause	43
— Tonsillitis, as cause	44
— Type of Diabetes Mellitus	79
— Peritonitis in	124
African Races, apparent immunity of	27
Age	28, 206
Albuminuria	88, 136
Alcohol, Intolerance of	215
— Tolerance of	215
Alcoholic Intoxication	209
Alimentary System	210
Almond Cakes	184
Amblyopia	110, 116

	PAGE
America, Mortality from Diabetes in	25, 26
Ammonia, increase of, in Urine	162
Anæmia of Brain	55
Analysis of Urine	214
Aneurisms, Capillary	116
— Aortic	208
Angeiomata	65
Angina Pectoris	120
Animal Diet	184
Anthrax causing Glycosuria	35
Antipyrin	199, 201, 217
Aortic Aneurism	208
Apoplectic Attacks	105, 108
Appetite	80
Arsenite of Bromine	182, 193, 198
Ascitic Effusion	124
Atrophy and Pigmentation of Inferior Ganglion of Vagus	59
— of Semilunar Ganglia	59
— of Nerve Cells	59
— of Pancreas	66
— of Optic Nerve	116
— of Gums	121
— of Stomach	124
Australia, Mortality from Diabetes Mellitus in	21

BABINGTON. p. 80; Bamberger, 124; Banatvala, 199; Barling, 64, 146; Barnes, 109; Baumann, 4; Baumel, 67, 68; Beale, 66, 210; Begbie (Warburton), 117; Bennett (Hughes), 43, 208; Bennewitz, 48; Bernard, 3, 7, 8, 9, 205; Bertillon,

INDEX.

20; Binz, 145; Blanc, 108; Bond, 135; Bordier, 126; Bouchard, 109; Bouchardat, 3, 68, 215; Brault, 65, 68; Brieger, 148: Bristowe, 99; Brücke, 6; Buhl, 157; Bull, 67, 143; Burdel, 45; Burge, 24; Burton, 13; Butel, 84; Butte, 10, 12, 15, 58, 59, 60; Buzzard, 107.

	PAGE
Bahamas, Mortality from Diabetes Mellitus in	21
Barbadoes, Mortality from Diabetes Mellitus in	22
Baths	192, 218
Beer, as cause	81
Belladonna with Opium	194
— with Ergot	217
Berkshire, Mortality from Diabetes Mellitus in	26
Berlin, Mortality from Diabetes Mellitus in	20
Bermuda, Mortality from Diabetes Mellitus in	21
Beta-crotonic Acid	88
Beta-oxybutyric Acid	88, 144
Bilateral Sciatica	105
Bladder, dilatation and Hypertrophy of	74
— Hæmorrhages in	74
— Irritable	215
Blisters	217
Blood	62, 82, 209
— Fatty	62
— Fat Embolism	62
— Acetone in	63
Body Weight, fluctuations of	— 82
Boils	36, 94, 215
Boracic Acid	193, 202
Borax	203
Bourboule Mineral Water	191
Bowels, Regulation of	202
Bradshawe Lecture	54
Brain	55
— Œdema of	55
— Anæmia of	55
— Tumours of	55
— Softening of	56
Corpora Amylacea in	56

	PAGE
Brain, Colloid Masses in	56
— Sclerosis of	56
— Alterations in Colour and Congestion of	56
— Enlargement of Perivascular spaces in	56
— Congestion of Medulla	56
— Cysts in white matter of	56
— Cysts in Choroid Plexus	56
— Dilatation of Ventricles	57
— Hæmorrhage into Brain substance	57
— Glycogen in Medulla and in vessels of Cerebral Cortex	57
— Iron in	57
Bread	185, 187
Breath, Odour of the	81, 136, 138, 150, 162
British Guiana, Mortality from Diabetes Mellitus in	21, 31
Bromide of Potassium	33, 193, 198
Bromine, Salts of	193
Buckinghamshire, Mortality from Diabetes in	26
Buckwheat Flour	187
Buenos Ayres, Incidence of Diabetes in	26
Burning of Palms of Hands	98

CALMETTI, p. 45; Campbell, 33; Cantani, 16, 68, 139; Cauldwell, 199; Cayala, 126; Charcot, 31, 135, 216; Chevers, 23, 30; Christie, 31; Clay, 208; Clubbe, 217; Colbourne, 26; Coldstream, 27; Cornillon, 105; Cornish, 23; Crooke, 75; Cullen, 3, 205; Cyon, 10; Cyr, 67, 135, 157; Czapek, 88.

Calculi, Pancreatic	69
California, Mortality from Diabetes Mellitus in	26
Cambridgeshire, Mortality from Diabetes Mellitus in	26
Camphor	199

LECTURES ON DIABETES.

	PAGE
Cape Town, Mortality from Diabetes Mellitus in	22
Capillary Aneurisms	116
Carbonate of Soda	199
Carbuncles	36, 94
Carcinoma of pineal gland	209
Cardiac Enlargement	61
— debility	62
Carlsbad Mineral Water	191
Cataract, Diabetic	110
Catarrh of Intestines	71
— of Stomach	71
— Pulmonary	117
Cats, Coma in	145
Causes of Diabetes 32, 206, 207
Cellulitis	100
Central Canal, Dilatation of	58
Cerebro-spinal Nerves	58
— Atrophy and Pigmentation of inferior Ganglion of Vagus	58
— Neuritis of central end of Vagus	58
— Tumours on Vagus Nerve	58
— Diabetic Neuritis	59
Ceylon, Mortality from Diabetes Mellitus in	21, 22, 81
Chili, Incidence of Diabetes in	26
China, Rarity of Diabetes in	24
Chloral Hydrate	201
Chloroform causing Glycosuria	33
Christiania, Mortality from Diabetes Mellitus in	20
Chronic type of Diabetes Mellitus	79
Cider, as cause	31
Circulatory System	60, 120
Cirrhosis of Liver	64, 124
— of Pancreas	69
Clemens' Solution	198
Climacteric Diabetes	49
Cocaine	199, 201
Codeine	193, 194
Cœliac Plexus	70
— Result of Extirpation of	60, 70

	PAGE
Cold, as cause	42, 208
— Climates, as cause	42
— Exposure to	185
Colloid masses in Brain	56
Coma Carcinomatosum	139
— Diabetic	108, 134
— Clinical History of	135
— Etiology of	135
— Symptoms of	136
— Pathology of	64, 188
— Küssmaul's	139
— Theories of	139
— Latham's theory of	145
— Cases of	147, 151, 158
— Causes of	161
— Temperature in	158
— Diagnosis of	160
— Prognosis of	163
— Treatment of	166
— Conclusions respecting	176
Complications	93
Congestion of Brain	56
— of Lungs	63
— of Liver	64
— of Stomach	70
— of Intestines	71
— of Kidneys	71
Constipation	135
Contagion causing Diabetes	33
Contusions	207
Copenhagen, Mortality from Diabetes Mellitus in	20
Corpora Amylacea in Brain	56
Creasote	199
Curare Diabetes	14
Cure, spontaneous	216
Cyprus, Incidence of Diabetes Mellitus in	21, 25
Cystic Disease of Pancreas	67, 208
Cysts of Brain	56

DALE, p. 84; Dalton, 7; Dastre, 8; Davenport, 49; Deane, 28; Debove, 33; Decker, 48; Defresne, 201; Desgranges, 216; Deutschmann, 110; Dickinson, 50, 56, 57,

INDEX.

65, 169, 202, 209, 210, 216; Dobson, 2; Donkin, 189; Dornblüth, 199; Draper, 199; Dreschfeld, 63, 135, 146; Dreyfous, 33, 109; Drummond, 207; Duffey, 67; Dufour, 6; Dujardin-Beaumetz, 185, 187, 199; Duncan, 59, 110, 207; Duncan (Matthews), 48, 208.

	PAGE
Deafness	116
Denmark, Mortality from Diabetes Mellitus in	20
Diabète Maigre	66
— Gras	67
Diabetes Insipidus	205
— Definition of	205
— Etiology of	205
— Predisposing causes of	206
— Morbid Anatomy of	209
— Pathology of	210
Symptoms and Clinical History of	211
General Health in	212
Prognosis of	216
Treatment of	217
— Mellitus	79
— Due to Muscular strain	38
— Pancreas in	66, 67
— Quantity of Urine in	84
— Types of	79
— Acute type	79
— Chronic type	79
— Sudden Death in	137
— Treatment of	181
— Œdema of	202
Diabetic Cataract	110
— Coma	134
— Liver	66
— Neuritis	59, 98
Diarrhœa	123
Dietary for Diabetics	135, 181, 191
Digestive Disturbance, as cause	32
— System	64, 121, 124
Dilatation of Ventricles of Brain	57
— of Bladder	74
— of Central Canal	58

	PAGE
Dilatation of Stomach	70
Diphtheria, as cause	44
Diplopia	110
Distension of Stomach	71, 151
Dogs, Diabetes in	27
Dover's Powder	203
Dresden, Mortality from Diabetes Mellitus in	20
Dropsy	102
Drowsiness	162
Drugs	193
Dry Mouth	215
Dryness of the Skin	215
Duodenum, Hæmorrhages in	71
Dupuytren's Contraction	126
Duration of Diabetes	92, 216
Dysenteric condition of Large Intestine	71
Dyspnœa	150

EAGER, p. 199; Ebstein, 34, 42, 71, 73, 124, 146, 158; Eckhard, 10, 60, 205; Edmunds, 116; Edwards, 42, 123; Ehrlich, 16, 72, 73; Eichhorst, 31, 217; Esterle, 208.

	PAGE
Ear Affections	116
Eczema of the Genitals	98
Egypt, Incidence of Diabetes in	25
Embolism, Fat	62, 64, 145, 149
Empyema	64
Endocarditis	61, 81, 120
England, Mortality from Diabetes Mellitus in	21
Enlargement of Peri-vascular Spaces	56
— of Peri-vascular Sheaths	58
— of Semilunar Ganglia and Splanchnic Nerves	59
— of Heart	61
— of Liver	64, 81, 124
— of Mesenteric Glands	71
— of Kidneys	71
Enteric Fever, as cause	46
— Prognosis of	124
Ephemeral Glycosuria	35
Epigastric pain	162
Epileptic attacks	105, 108

Ergot 217	Fehling's Solution ... 88
— with Belladonna ... 217	Ferric Chloride reaction ... 91, 140, 152, 162
Ergotin 217	
Erysipelas causing Glycosuria 85	Fibroid, Uterine 208
	Fingers, Painful Affection of 107
Erythema 94, 97	
Erythematous Œdema ... 97	Florador 186
Etiology 20	Fluctuations of Body Weight 82
Eugenia Jambolana ... 199	
Europe, Mortality from Diabetes Mellitus in ... 20	Fractures Subcutaneous, causing Glycosuria ... 85
Eustachian Tube, œdematous swelling of ... 116	France, Incidence of Diabetes in 27
Exalgine 199, 201	Furious Mania 108
Excessive sexual indulgence, as cause ... 50	GAILLARD, *pp.* 65, 69; Galen, 2; Galezowski, 109, 216; Gamgee, 163; Gardner, 199; Garnerus. 216; Gaucher, 33; Gemmell, 199, 201; Gerard. 88; Gerhardt, 124, 140; Gerrard, 89; Gowers, 113, 116, 216; Graefe. 110; Graham, 24; Grant Bey, 25; Gregory, 8; Grenier, 43; Gresham, 26; Griesinger, 116, 124; Griswold. 49; Guinon, 109; Gull. 99.
Exciting causes ... 32, 207	
Exophthalmic Goitre and Diabetes ... 43, 109	
Extirpation of Cœliac Plexus, result of ... 60	
— of Pancreas, result of 66	
Eye Affections ... 109, 216	
FAGGE, *pp.* 27, 79, 103, 166, 215; Fehling, 5; Fenn. 24; Ferraro, 27, 71; Fichtner, 71; Finlayson, 108; Fienzal, 112; Finkler, 192; Fischer. 167; Flatten, 207; Fleischer, 139, 140; Florain, 107; Fort, 83; Foster (B.W.), 16, 62, 135, 166; Foster (M.), 13, 16; Fowler, 83; Foxwell, 148; France, 110; Frank, 208; Frew, 30; Frerichs, 15, 27, 36. 56, 58, 62, 66, 71, 137, 146, 150, 158, 166, 189; Frison, 67; Fütterer, 57.	
	Galvanisation 217
	Gangræna Bullosa Serpiginosa 101
	Gangrene of Lungs 63. 64. 119
	— of Kidney 71
	— of both lower Extremities 100
	— of Toes 101
	Gastric Juice 124
	Gastritis, Interstitial ... 124
	General Peripheral Neuritis 107
Fat 191	Genitals, Eczema of ... 98
— Embolism 62, 64, 145, 149	Genito-urinary System ... 71
Fatty Heart 61, 62, 81, 120, 146	Geographical Distribution 20
— Blood 62	Gibraltar, Mortality from Diabetes Mellitus in ... 21
— Liver 64, 65	
-- Kidney 71, 74	Giddiness 105
Febrile Diseases ... 208	Glucose 4
Feebleness of Pulse ... 162	Gluten Bread 185
Feet, Hyperæsthesia of Soles of 107	Glycerine 188, 192

	PAGE
Glycerinum Acidi Carbolici	202
Glycocholic Acid	7
Glycocin	7
Glycogen, Preparation of	5
— Quantity of in Liver	6
— Distribution of	6
— Formation of	6, 7
— Conversion of	7
— in Medulla and in vessels of Cerebral Cortex	57
— in Heart	62
— in Spleen	66
— in Urine	85
— in Kidney	78
Glycogenic Degeneration of Heart	147, 150
Glycosuria	4
— Experimental	9
— Pancreatic	10
— Toxic	13, 32
— with Jaundice	15
— Theories of	15
— at La Trappe	31, 135
— from Digestive Disturbance	32
— from Nervous Disturbance	33
— Ephemeral	35
— from Gangrene	35
— from Malignant Disease of Abdomen	36
— Acute Febrile	80
— in Healthy Persons	83
— in Disease	83
— Intermittent	85
Glycuronic Acid	5
Goitre, Exophthalmic	43, 109
Gout, as cause	43
Gums, Affections of	121

HAGER p. 198; Hallervorden, 85; Hamilton, 16, 42, 64, 145, 156; Hanot, 64; Harker, 92; Harley, 58; Harrison, 85; Harvey, 26; Hassall, 186; Hassenauer, 25; Hay, 13; Heller, 124; Henrot, 58; Hensen, 10, 60; Hermanides, 42;

	PAGE
Heynsius, 7; Hirsch, 2; Hirschman, 113; Honigmann, 124; Hoppe-Seyler, 136; Haughton, 208; Heusoner, 101; Hull, 44; Hunt, 100.	
Hæmocytes, Destruction of	64
Hæmorrhage into Brain substance	57
Hæmorrhages in Stomach	70
— in Duodenum	71
— in Bladder	74
— in Retina	113
Hæmorrhagic Infarcts in Lungs	63
Hands, Burning of Palms of	98
Headache	105
Hearing	80
Heart	60
— Dilated or Hypertrophied	61, 62
— Fatty	61, 62, 120, 146
— Pericarditis	61
— Fatty Blood in	61
— Valvular Disease of	61, 120
— Endocarditis	61, 120
— Debility of	62
— Glycogenic Degeneration of	147, 150
— Glycogenic Deposits in	62
— Affections of Wall of	120
— Fibroid Degeneration of	120
— Failure	146
Heligoland, Mortality from Diabetes Mellitus in	21
Hemiplegia	108
Hepatitis, Interstitial	64
Heredity	30, 206
Hippuric Acid in Urine	88
History	2
Hongkong, Mortality from Diabetes Mellitus in	21
Hot Sun	208
Hot Water Baths	218
Hunger	215
Hyaline Degeneration of Tubular Epithelium	71
Hyperæsthesia of the Soles of the Feet	107
Hypertrophy of Heart	61, 62

	PAGE
Hypertrophy of Bladder	74
Hyposulphite of Soda ...	202
Hysteria and Diabetes ...	43
IMLACH, *p.* 50; Israel, 74.	
India, Incidence of Diabetes in... 23,	31
Indican in Urine... ...	88
Inflammatory Diseases, predisposition to 134,	208
Influenza, as cause ...	44
Inhalations	167
Injections... ... 167,	202
Injuries causing Diabetes	34
Inosite 4,	215
Insanity	30
Integumentary System ...	93
Intermittent Glycosuria	85
— Polyuria	215
Interstitial Gastritis ...	124
Intestines	71
— Hardened fæcal masses in	71
— Congestion or Catarrhal conditions of Mucous Membrane	71
— Hæmorrhages in Duodenum	71
— Mesenteric Glands enlarged	71
— Dysenteric condition of Large Intestine ...	71
— Ulceration of	123
Intolerance of Alcohol ...	215
Iodide of Potassium ...	217
Iodoform 197,	203
Ireland, Mortality from Diabetes Mellitus in ...21,	27
Iritis	113
Iron in Brain	57
— in Liver	66
— in Spleen	66
Irritability of Temper ...	80
Irritable Bladder ...	215
Italy, Incidence of Diabetes Mellitus in ...21,	27

JACOBI, *p.* 209; Jaeger, 115; Jarrold, 209; Jensen, 116; Johnson, 90; Jones (Bence), 166; Jones (Handfield), 67; Jordão, 116.

	PAGE
Jamaica, Mortality from Diabetes Mellitus in ...	21
Jambul 35,	199
— Perles	200
Jaundice and Glycosuria 15,	124
Jews, frequency of Diabetes among	27

KAPOSI, *pp.* 98, 101, 102; Kaulich, 139; Kennedy, 199, 217; Kinsey, 22; Kingsbury, 199; Kirchner, 117; Kirk, 14; Kirmissen, 101; Kisch, 31; Klebs, 10, 59, 60; Kratschmer, 31; Kühne, 7; Külz, 6, 15, 116, 212, 216; Küssmaul, 134. 138, 142, 145, 146, 150, 162, 166, 192; Küster, 208.

Keratitis	113
Kidneys	71
— Fatty Degeneration of 71,	74
— Enlargement and Congestion of	71
— Tubercle of	71
— Lardaceous Disease of	71
— Gangrene of	71
— Hyaline Degeneration of Tubular Epithelium	71
— Chronic Diffuse Nephritis	74
— Glycogen in	73
Knee-jerks, loss of 80.	108
Küssmaul's Coma ...	139

LABBÉ, *p.* 33; Lacombe, 207, 216; Lancereaux, 67; Landois, 5, 7; Langenhans, 67; Lannois, 12; Laplane, 217; Latham, 3,16, 145; Laycock, 216; Leber, 113; Leconte, 13; Lecorché, 33, 49, 61, 120; Legg (Wickham), 6, 15; Legrand du Saulle, 105; Le Nobel, 88, 91, 140, 142; Lépine, 8, 11, 12, 66, 68, 108, 135, 136, 157, 201; Leroux, 97;

INDEX. 227

Letulle, 33, 64; Leube, 85, 104; Leyden, 59; Libby, 217; Lindsay, 34; Litten, 139, 142; Longstreth, 67; Lowinsky, 215; Lubimoff, 58, 59; Lunin, 209; Lustig, 60; Luys, 56.
Lævulose 4, 85
Lactose 4
Lancashire, Mortality from Diabetes in 27
Lardaceous Disease of Kidney 71
Large Intestine, Dysenteric condition of ... 71
Larynx and Trachea, Membranous Inflammation of 117
Latham's theory of Diabetic Coma 145
La Trappe, Glycosuria at 31, 135
Leipzig, Mortality from Diabetis Mellitus in ... 20
Lightning Stroke, as cause 42
Lincolnshire, Mortality from Diabetes in ... 26
Lipæmia 62
— and Fat Embolism ... 145
Liquor Clementis ... 198
— Ouabaïni 201
Lithia 198
Liver 64
— Diseases and Diabetes 44
— Sugar 9
— Enlargement of 64, 81, 124
— Small, Pale, and Soft 64
— Fatty Degeneration of 64
— Congestion of 64
— Interstitial Hepatitis 64
— Cirrhosis of ... 64, 124
— Destruction of Hæmocytes 64
— Abscess of 65
— Thrombosis of Portal Vein 65
— Angeiomata 65
— Fat in 66
— Iron in 66
— Glycogen in 6
London, Mortality from Diabetes Mellitus in ... 20

Long Journeys, Danger of 134, 135, 161
Loss of Knee-jerk ... 108
— of Sexual desire ... 107
Lungs 63
— Congestion and Œdema of 63
— Hæmorrhagic Infarcts in 63
— Gangrene of 63, 64
— Pneumonia 63, 64
— Phthisis 63, 64
— Pleurisy and Empyema 64
— Fat Embolisms ... 64
— Affections of 81
— Softening of 63
Lymphangeitis causing Glycosuria 85

MACKENZIE, pp. 55, 72, 110, 135, 137, 143; Mac Munn, 148; Macnamara, 26; Madigan, 105; Maguire, 61, 88, 120; Mauby, 43; Mann, 161; Marcet, 98; Marchal (de Calvi), 99; Marie, 109; Marsh, 42; Marshall, 42; Masoni, 14; Maudsley, 42; Max Einhorn, 88; Mayer, 61; McAvoy, 202; McBride, 88; Meyer, 5; Minkowski, 10, 11, 66, 144, 161; Minot, 157; Miot, 116; M'Nish, 36; Moleschott, 199; Monckton, 199; Moore (Norman), 67, 90; Moosdorf, 34; Müller, 85; Munk, 59, 89; Murrell, 217.
Madagascar, Incidence of Diabetes in 24
Malaria, as cause... ... 45
Malignant Disease ... 36
Malta, Mortality from Diabetes Mellitus in ... 21
Mania, Temporary ... 105
— Furious 108
Martineau's Specific ... 198
Massage 192, 202

228 LECTURES ON DIABETES.

	PAGE
Masturbation, as cause ...	50
Mauritius, Mortality from Diabetes Mellitus in ...	21, 81
Medulla, Congestion of ...	56
Melancholia	105
Membranous Inflammation of Larynx and Trachea...	117
Menstruation	80
Mental Emotion, risk of	134, 208
Menthol	201
Mercury	217
Mesenteric Glands, Enlargement of	71
Mongolian Races, apparent immunity of	27
Montserrat, Mortality from Diabetes Mellitus in ...	21
Morbid Anatomy... ...	54
Morphia	201
Mortality from Diabetes Mellitus in British Colonies	21
— in Europe	20
Mouth, condition of	128, 215
Multiple Pancreatic Abscesses	67
Muscular effort, as cause	38, 135, 208

NEDATS (DE), p, 187; Nettleship, 110, 116; Neuffer, 210; Norris, 155; Nothnägel, 207.	
Nails	80
Naples, Mortality from Diabetes Mellitus in ...	20
Natal, Mortality from Diabetes Mellitus in ...	21
Nephritis, chronic diffuse	74
Nervous Disturbance, as cause	38, 42, 135
Nervous System in Diabetes	55, 104, 209
Neuralgia	105, 201
Neuritis of Central end of Vagus	58
— Diabetic	59, 98
— Peripheral	107
New Growths	129

	PAGE
New South Wales, Mortality from Diabetes Mellitus in	21
New York, Mortality from Diabetes Mellitus in ...	25
New Zealand, Mortality from Diabetes Mellitus in	21
Nitric Acid ...	199, 217
Nitro-glycerine	199
Nitro-hydrochloric Acid	199
Nitro-muriatic Acid ...	217
Normandy, Incidence of Diabetes in ...	27, 81
Norway, Mortality from Diabetes Mellitus in ...	21
Nux Vomica	199

O'DONNELL, 187; Oliver, 104, 199, 201, 207; Opitz, 217; Ord, 48; Otto, 88.	
Obesity	81, 212
Odour of the Breath	81, 136, 138, 150, 162
Œdema of Brain ...	55
— of Diabetes	202
— of Lungs	63
— of Trunk and Lower Limbs	102, 215
— Erythematous ...	97
Œdematous Swelling of Eustachian Tube ...	116
Oil of Peppermint ...	203
Opium ...	88, 193, 194, 201
— with Belladonna ...	194
Optic Nerve, Atrophy of	116
Otitis Media	116
Oxalic Acid in Urine ...	88
Oxfordshire, Mortality from Diabetes in ...	26
Ouabaïn	199, 201
Ouabaïs Plant	201
Ovaries	74
— Gangrene of	74

PAVY, pp. 3, 7, 10, 13, 28, 29, 34, 60, 93, 104, 105, 128; Peiper, 42, 60; Penzoldt, 143; Percy, 59; Petters, 139; Peyrani, 206; Peyraud, 199;

INDEX. 229

Pisenti, 73; Plumert, 124; Poniklo, 42; Ponomaroff, 124; Popper, 68; Portcous, 189; Poulet, 48; Pribram, 215, 216; Prout, 27, 42, 79, 134, 161; Pryce (Davies), 97.
Painful Affection of the Fingers 107
Pancreas 66
— Extirpation of ... 66
— Atrophy of 66
— Fibroid 66
— Enlarged 67
— Cystic Disease of 67, 208
— Multiple Abscesses of 67
— Theories of Action of 67
— Pigmentary Cirrhosis of 69
— Calculi 69
Pancreatic Glycosuria ... 10
Pancreatine 201
Papillomatosis Diabetica 98
Paraglucose 16
Paralysis 107
— of a Limb 107
Paralytic Distension of Stomach ... 151
Paraplegia 107
Parenchymatous Retinitis 113
Paris, Mortality from Diabetes Mellitus in ...20, 21
Pepsin 199
Peptone, Conversion of into Sugar 9
Perchloride of Iron reaction 136
Perforating Ulcer ... 101
Peripheral Neuritis ... 107
Peritonitis, Acute ... 124
Peri-vascular Sheaths, Enlargement of ... 58
— Spaces, Enlargement of 56
Persia, Incidence of Diabetes in 25, 31
Phenacetin 199
Philadelphia, Mortality from Diabetes in ... 25
Phlegmonous Affections causing Glycosuria ... 35
Phloridzin 13
Phosphorus ... 199, 201

Phthisis ... 31, 63, 64, 117
Pilocarpine ...199, 201, 202
Pineal Gland, Carcinoma of 209
Pleurisy 64
Pneumaturia 85
Pneumonia 63
Pneumonic Phthisis, Chronic 117
Polyuria, Intermittent ... 215
Portal Vein, Thrombosis of 65
Potatoes as Food 187
— Influence of 188
Predisposing causes ... 20
Pregnancy, as cause 48, 208
Prevalence of Diabetes ... 27
Prognosis 92
Pruritus Vulvæ ... 90, 202
Prussia, Mortality from Diabetes Mellitus in 21, 31
Ptosis 107
Pulse 81, 136, 162
Purpura 98
Pyrexia resembling Typhoid 125

QUANJER, p. 199; Quincke, 66, 163.
Quantity of Urine in Diabetes 84
Queensland, Mortality from Diabetes Mellitus in 21

RALFE, pp. 91, 189, 208; Ratcliffe, 75; Rayer, 104, 124; Raynaud, 116, 117; Redard, 35; Reudu, 33; Renzi, 36; Reumont, 42; Richardière, 42; Rickards, 135; Riess, 139; Roberts (Prescott), 158; Roberts (W.), 28, 89, 102, 189, 210, 216, 217; Robertson, 27; Roger, 44; Rogers, 44; Rollo, 3, 42, 82, 110, 123; Rosenbach, 13; Rosenblath, 99; Rosenstein, 109, 124; Routh, 203; Rupstein, 139, 140; Ryba, 124.

	PAGE
Race	27
Rapid increase of Diabetes	28
Rapidity of Pulse as a prodromal sign of Coma	136, 162
Registrar-General's Report	29
Regulation of the Bowels	202
Respiration, altered	162
Respiratory System	63, 117, 210
Retinal Affections	113
Rheumatism	43, 126
Rome, Mortality from Diabetes Mellitus in	20
Royat Mineral Water	191
Rutland, Mortality from Diabetes in	26

SALKOWSKI, p. 140; Salomon, 143; Salomonson, 107; Samelsohn, 110, 113, 116; Sanders, 64, 145, 156; Saundby, 64; Savage, 42; Schachman, 64; Scherer, 215; Scheuplein, 34; Schiff, 10, 13, 58; Schmiedeberg, 5; Schmitz, 28, 29, 33, 42, 61, 92, 146, 203; Schapiro, 209, 210; Schulzen, 215; Scudamore, 27; Sée (Germain), 199; Seegen, 7, 8, 16, 31, 63, 82, 85; Seifert, 124; Senator, 84, 136, 138, 144, 157, 160, 205, 208, 215, 216; Simon, 98, 104; Sinclair, 49; Skerritt, 80; Slater, 85; Smith (Shingleton), 42, 58, 59; Spencer, 36, 102, 110; Squire (Balmanno), 199, 201; Stadelmann, 85, 162, 199; Stern, 30; Stockvis, 73, 146; Strahan, 107; Strange, 205; Strauss, 72, 73, 209, 215; Strumpell, 108; Sydenham, 2.

Saccharin	192
Salicin	193, 199
Salicylate of Sodium	193, 198
Salicylic Acid	193

	PAGE
Salts of Bromine	193
Salutaris Water	191
Sciatica, Bilateral	105
— Unilateral	106
Sclerosis of Brain	56
— of Sympathetic Ganglia	59
Scotland, Mortality from Diabetes Mellitus in	21, 27
Scotoma, Central	116
Semilunar Ganglia	149
— Enlargement of	59
— Atrophy of	59
Semolina	186
Septicæmia	36
Sex	30
Sexual Continence, as cause	50
— Indulgence, as cause	50
— Power	80, 107
Sierra Leone, Mortality from Diabetes Mellitus in	21
Singapore, Incidence of Diabetes in	24
Skatoxylsulphuric Acid in Urine	88
Skim Milk	189
Skin	80
— Dryness of	215
Smell	80, 116
Sodium Butyrate, effect of on cats	145
Softening of Brain	56
— of Spinal Cord	58
— of Lungs	63
— of Stomach	70
Source Victor Mineral Water	191
Soya Bread	185
Spinal Cord	57
— Dilatation of Central Canal	58
— Enlargement of Perivascular Sheaths	58
— Localised Softening of	58
— Tumours of	58
Splanchnic Nerves, Enlargement of	59
Spleen	66
— Iron in	66
— Glycogen in	66
Spontaneous Cure	216

INDEX.

Staffordshire, Mortality from Diabetes in ... 27
Starchy Food, excessive use of 31
Steam Baths 218
St. Helena, Mortality from Diabetes Mellitus in ... 21
St. Kitts, Mortality from Diabetes Mellitus in ... 21
Stomach 70
— Hæmorrhages in ... 70
— Congestion of 70
— Softening of 70
— Dilatation of 70
— Chronic Catarrh of ... 71
Thickening of Mucous Membrane of ... 71
— Distension of ... 71, 151
— Sense of Emptiness at pit of 123
— Ulceration of 123
— Atrophy of 124
Stools 81
Strabismus 107
Subcutaneous Fractures causing Glycosuria ... 35
Substitute for Bread ... 187
Sudden Death in Diabetes 137
Suffolk, Mortality from Diabetes in 26
Sugar and Starchy Food, excessive use of ... 31
Sugar Ferment 8
— in Liver 9
— in Urine 85
— Tests for 88
— Production 7, 9
Disappearance of ... 104
Suicidal tendencies ... 105
Sulphide of Calcium ... 199
Sulpho-carbolate of Soda 199
Sulphuret of Potassium... 202
Sussex, Mortality from Diabetes in 26
Sweating 98, 203
Symmetrical Erythema... 97
Sympathetic Nerves ... 59
— Thickening of Semilunar Ganglia and Splanchnic Nerves 59
— Sclerosis of Sympathetic Ganglia ... 59
Sympathetic Atrophy of Semilunar Ganglia ... 59
— Atrophy of Nerve Cells 59
— Result of Extirpation of Cœliac Plexus ... 60
Symptoms, Treatment of 201
Syphilis, as cause 47, 208
Syzygium Jambolanum... 199

Tait, pp. 50, 202; Tardieu, 56; Taylor, 135, 137, 156, 166; Teissier, 199; Teschemacher, 85; Thierfelder, 13; Tholozan, 25, 27, 31; Tilden, 157; Toynbee, 116; Trommer, 5; Trousseau, 20, 31, 110, 212, 215; Tuffier, 129; Turner, 71; Tyson, 25.
Tasmania, Mortality from Diabetes Mellitus in ... 21
Taste 80, 116
Teeth 80
Temperature ... 80, 103, 158
Temporary Mania ... 105
Test for Acetone... ... 91
— for Sugar 88
Thickening of Mucous Membrane of Stomach 71
Thirst 201
Thrombosis of Portal Vein 65
Tinnitus 116
Tobacco 116
Toes, Gangrene of ... 101
Tolerance of Alcohol ... 215
Tongue 80
Tonsillitis, as cause ... 44
Toxic Glycosuria ...13, 32
Travelling, great risks in 134, 161
Treatment of Diabetes Mellitus 181
— Insipidus 217
— of Symptoms 201
Tubercle of Kidney ... 71
Tubercular Diathesis ... 208
Tumours of Brain ... 55
— of Spinal Cord ... 58
— on Vagus Nerve ... 58
— of Abdomen 208

	PAGE
Turks, Prevalence of Diabetes among	27
Turpentine	217
Tuscany, Incidence of Diabetes in	27
Types of Diabetes Mellitus	79
Typhoid, Intercurrent Pyrexia resembling	125
Ulcer, Perforating	101
Ulceration of Stomach and Intestines	128
Unilateral Sciatica	106
United States, Mortality from Diabetes in	25
Uræmia	146
Uranium Nitrate	193
Urinary System in Diabetes Insipidus	210
Urine	82
— Quantity of in Diabetes	84
— Specific gravity of	84
— Colour	84
— Reaction	84
— Deposits	84
— Changes in	84
— Normal Constituents	84
— Abnormal Substances in	85
Increase of in Diabetes Insipidus	215
Uterine Fibroid	208

VALENTINE, *p.* 199; Vergeley, 120; Villemin, 194; Von Jaksch, 62, 136, 139, 140; Von Kormick, 13; Von Mering, 10, 14; Von Wittich, 6, 15; Voss, 212.

	PAGE
Vagus Nerve, affections of	58
Valerian	217
Valerianate of Zinc	217
Vals Mineral Water	191
Valvular Disease	61, 120
Vaseline	201
Vaso-motor Theory	16
Vichy Water	191
Victoria, Mortality from Diabetes Mellitus in	21
Vienna, Mortality from Diabetes Mellitus in	20
Vision	80

WALTER, *p.* 144; Watson, 102, 105; Weber, 33; Wedenski, 4; Weil, 85, 206, 208, 216; Westphal, 208; Weyl, 65; White (Hale), 43, 59; Wilde, 110; Wilks, 199; Willan, 110; Willis, 2, 205; Windle, 55, 57, 61, 71, 135, 136, 153, 154; Worms, 93; Wyatt, 15.

Warwickshire, Mortality from Diabetes in	27
Waters, Mineral	191
White Spots on Retina	113
Xanthoma Tuberosum	99

YARROW, *p.*50; Young, 112.

Yorkshire, Mortality from Diabetes in	27

ZALESKI, *p.* 57, 66; Zenner, 217; Ziemssen, 16; Zinn, 44.

Zinc Ointment	202

RECENT MEDICAL WORKS
PUBLISHED BY JOHN WRIGHT & Co., BRISTOL.

BY THE SAME AUTHOR.

8vo. 300 pp., with 50 original Illustrations. Price 6/6, post free.

LECTURES ON BRIGHT'S DISEASE.

EXTRACTS FROM PRESS NOTICES.

"The task of writing a new treatise on a subject which, in spite of the numerous researches made since Bright published his reports in 1827, has still many obscure points in its pathology and pathogenesis, is by no means an easy one. Dr. Saundby, however, has accomplished this task very ably, and has succeeded in putting together, in less than 300 pages, an admirable account of Bright's disease in its various forms. It is so thoroughly practical that it recommends itself to the busy practitioner, and yet so scientific that even the specialist may consult its pages with great profit. Dr. Saundby's lectures are not merely compilations; the author has himself materially assisted in the unravelling of some of the many obscure points in the pathology of Bright's disease, and has a large clinical material to draw upon for the illustration of his lectures. . . . Though the author's classification may be objected to, the description of the symptoms in the several cases which are given for illustration leaves nothing to be desired, and the same may be said of the chapter on treatment with which the work concludes. Dr. Saundby has succeeded in giving, in a clear and concise form, an accurate account of our present knowledge on albuminuria and Bright's disease. The work is written in an easy and readable style, and the illustrations are numerous and well executed."—*British Medical Journal.*

"A most important and valuable contribution to Bright's disease has been published during the year by Dr. Saundby, whose lectures handle the subjects of albuminuria, uræmia, renal dropsy, and the cardio-vascular changes in an admirable manner. The latest information on these matters will be found set before the reader of these chapters in a terse but very lucid style, while those devoted to the special pathology of the various forms of Bright's disease are enriched by the author's own extensive personal experience. It is a work which will be extensively read and referred to."—*British Medical Journal. Second Notice.*

"Dr. Saundby has long been known as a diligent and indefatigable worker in the field of renal pathology and therapeutics, so that these lectures will command for themselves a recognition beyond what any words of commendation can bestow. And, indeed, these lectures are full of good things, for, though not an "epoch-making" book, it is one of those works which throws a clear and steady light on doubtful and obscure questions, while arranging and classifying facts already accepted, but to which no recognised position had as yet been assigned. In other words, Dr. Saundby in these lectures has focussed for us the point to which our knowledge of the disease in question has reached, and placed that knowledge concisely and lucidly before us. . . We say that these lectures should be attentively read by all who are desirous of gaining a clear insight into what is meant by Bright's disease and who may wish to follow intelligently the various changes and complications likely to occur in any given case under their care."—*Lancet.*

"A monograph which is very opportune and especially worth reading, and which will win many friends among us."—*Intern. Klinische Rundschau.*

RECENT MEDICAL WORKS

Published by John Wright & Co., Bristol.

"We perused with enjoyment and profit this valuable work. . . . It presents in a small compact form the results of much careful work and hard thinking on a subject with which Dr. Saundby has peculiarly associated his name. We ought to add that a number of beautiful drawings illustrate the text."—*Practitioner.*

"It is an able and luminous essay on this oftentimes obscure and puzzling affection. . . . The section on the clinical examination of the urine is very full and complete, and, taken in connection with the rest of the work, it will be found of great value to all who are called upon to make an investigation into the condition of suspected samples of urine. . . . The chapter on treatment is a very useful and suggestive *resume*, which derives much advantage from the pains which Dr. Saundby has taken to embody in it the best results of his own large experience; and, indeed, the whole work gains largely in value and importance from this personal inspiration of its pages. We strongly commend it to the notice of practitioners, who will gather much from its attentive perusal."—*The London Medical Recorder.*

"The author studies with much method and clearness the essential phenomena of Bright's disease. . . . At the close of each chapter is found a bibliographical index, generally very complete, and which shows great learning on the part of the author. In short, this work, which does not contain 300 pages, and which is illustrated with numerous engravings, is remarkable for conciseness, clearness, and practical sense. It is the work of a sound clinician."—*Le Progres Medical.*

The author has long been known as an authority on everything pertaining to renal disease; and his views on renal pathology, and his methods of testing for albumen, are now accepted by those most competent to judge of such matters. . . . A marked and useful feature of this book is the copious bibliography which follows each chapter, and which is well brought up to date. Dr. Saundby manages, however, in each chapter to give all the pith of the most important papers, and thus saves the reader the trouble of referring to the papers themselves. Being, as is well known, well posted in foreign medical literature, especially German, he is able to furnish his readers with information from the original sources, not at second hand, as is but too often the case with text-book writers who sometimes are either ignorant of German or too lazy to take the trouble of verifying quotations. . . . This book meets a long-felt want, and will, we believe, have as large a circulation as it deserves. The type and paper are very good, and, when necessary, original illustrations are introduced into the text."—*Birmingham Medical Review.*

"This is a German translation of Dr. Saundby's Lectures on Bright's disease, published last year. It is so seldom that our German colleagues pay any English medical book the compliment of translating it, that the occurrence deserves notice. The translation appears to have been done well and carefully, the translator adding footnotes of his own, and in some cases not scrupling to advance arguments against the views expressed in the text. In his preface the translator says, 'I believe these lectures will not only be a useful guide to young practitioners, but that experienced physicians also will derive scientific and practical instruction from them; and they will not be without interest even to clinical teachers. Finally, I am convinced that this book is indispensable to every one who desires to work scientifically at the advancement of this department of medical knowledge.'"—*Birmingham Medical Review.* Second Notice.

"It is well known to the profession that Dr. Saundby has long been an earnest worker in the elucidation of questions connected with Bright's disease. We may say at once that we have read the book with great pleasure and profit, and WE VERY WARMLY RECOMMEND ITS STUDY BY THE PROFESSION. The size of the work is so modest (290 pages) that on that ground alone it invites perusal, and the volume recommends itself from the fact that it will easily find a nook on the library shelves. As might have been expected from

RECENT MEDICAL WORKS
PUBLISHED BY JOHN WRIGHT & CO., BRISTOL.

Dr. Saundby, the matter, taken both from personal observation and wide general reading, is accurate, well digested, and concisely and perspicuously arranged. For those who wish to follow up the various points discussed there is a bibliography given at the end of each chapter, and the text is illustrated by many excellent and useful figures. . . . Considering, therefore, the prevalence of the diseases discussed, their importance, and the hazy notions held by many regarding them, we believe that this little work will supply a great want, and will meet with the large measure of success which it deserves. We may add that the book is most excellently and carefully printed."—*Illustrated Medical News*.

" It is a book which no practitioner should be without, as though it is so concise, yet there is no subject connected with albuminuria on which the most recent information is not to be had, together with a very full bibliography of all recent works on each branch of the subject, which those desirous of fuller information may consult. This bibliography has not been constructed from a library catalogue, but every work has evidently been well read and studied by the author, and is so referred to in the text that the student will have no difficulty in selecting the treatises that will be useful to him in his inquiries. . . . There is no ambiguity in regard to Saundby's opinions, which are enunciated with as much clearness and distinctness as brevity. . . . It would be difficult to point out another work in which the history, classification, and etiology of albuminuria is comprised in fifteen pages; and yet no one, we think, could rise from the perusal of these fifteen pages without having acquired clearer and more distinct views of the different pathological theories that have been propounded to account for the various forms of kidney disease from the days of Bright down to the present time. And yet these theories are so concisely described that no one, not in his dotage, is ever likely again to forget them. . . . We have great pleasure in recommending this book as a most handy and useful one for the consulting room, one which may be picked up and referred to on any subject connected with the diagnosis and treatment of albuminuria, with the certainty of obtaining the most recent information regarding it. It is beautifully printed, and the illustrations are distinct and accurate. It is altogether a work of which both author and publisher may be proud as a work of art."—*Edinburgh Medical Journal*.

" We have no hesitation in pronouncing the volume a most excellent one, and in every way deserving the confidence of the profession."—*Glasgow Medical Journal*.

" The work Dr. Saundby has done in these lectures amplifies and diversifies the results of older discovery rather than unsettles the fundamental truths themselves. . . . It is fitting that, after years of careful observation and original enquiry, he should bring into focus his results, and present them to the medical profession in the form of a well-digested and scientifically arranged volume such as we have now before us. One remarkable fact which is everywhere obvious throughout this work is the infinite ramifications that may be worked out in a subject which appeared to have received its final touches years ago. . . . Respecting the functions and diseases of the kidney physiological chemistry is exploring to some purpose, and Dr. Saundby is just the man to give these results a clear and intelligible exposition. His chapters on Dropsy, the anatomy and pathology of the kidney, and on clinical symptoms and complications, are very fresh and original. . . . He does not give the three broad distinctions of Bright's disease in conformity with older writers, but judges from his own microscopical examinations that hard and fast lines cannot be drawn. . . . The book is a distinct advance on the previous literature of Bright's disease; it is well written, well illustrated, easily assimilated, and therefore to be commended as a work for the general medical profession." —*Glasgow Herald*.

" An exceedingly good review of the knowledge of Bright's disease."—*New York Medical Record*.

RECENT MEDICAL WORKS

Published by John Wright & Co., Bristol.

"This is an ably written book, bearing the stamp on nearly every page of a sound and independent judgment."—*Liverpool Medico-Chirurgical Journal.*

"A work which gives, within a reasonable compass, all that is best in modern literature, supported at the same time by individual practical experience, so that it is a work which the conscientious general practitioner ought to have by him. We must congratulate the publishers on the excellent way the book has been produced."—*Provincial Medical Journal.*

"No such succinct and comprehensive *resume* of the subject has ever appeared from any pen."—*Monthly Magazine of Pharmacy.*

"An original and most valuable contribution to the literature of kidney disease."—*Dublin Medical Journal.*

"It is a genuine pleasure to say that a book is a good one. . . . Without hesitation we can say Dr. Saundby's book is both solid and good. . . . Worth diligent study from cover to cover, and, except upon the one point of classification, there is nothing better extant in our language."—*Bristol Medical Chir. Journal.*

"We think it an excellent book. . . . The illustrations with which this work is plentifully supplied will moreover enhance its value. Altogether it is a work which can be confidently recommended to the student as well as the practitioner."—*Hosp. Gazette.*

"This excellent book is already well known and highly appreciated by the medical profession."—*Manchester Medical Chronicle.*

"Any new writer coming into this field can hardly claim much distinction unless he has new results to bring from practical experience. Dr. Saundby's lectures on Bright's disease do this; and would make a valuable book were it only for this reason. But they have further special merits in the ability with which the history, classification, and etiology of albuminuria are condensed and made concise for study or reference, without any scampering or hurrying, and in the extensive references to the best current books upon diseases of the kidneys that have been added in recent times to foreign medical literature. Such qualities make the work a good textbook for students. Medical men in every stage of development, however, will be quick to recognise the worth and authority of Dr. Saundby's own opinions, illustrations, and tests. His large experience and long continued study of his subject give the book more weight than attaches to a simply educational treatise."—*Scotsman.*

"In all the sections it may be said that Dr. Saundby's book is well written up to the times, and contains not only a summary of recent researches, but also the results of personal observations and experiments through a long series of years. . . . Dr. Saundby's work is a good one, and well worthy of a careful perusal."—*Australian Medical Journal.*

"This volume of Dr. Saundby's is a valuable contribution. He has a clear grasp of the facts that have passed before him during many years, and presents his conclusions in a concise and satisfactory manner. The whole subject of Bright's disease receives attention; the present state of our knowledge is finely stated, and the author's own experience carefully detailed. The work will be found to repay diligent study. The illustrations are very good."—*Buffalo Medical and Surgical Journal.*

"The subject is very thoroughly investigated, comprising extensive clinical experience and the gist of contemporary authorities."—*The Sanitarian, New York.*

"The work is done in a very practical manner. It is perhaps one of the best works on the subject, and is presented in a form to be comprehended. Good paper, type, and binding."—*Physio-Medical Journal, Indianapolis.*

RECENT MEDICAL WORKS

Published by John Wright & Co., Bristol.

"A concise and interesting volume, and deserves to become popular in the bibliography of this important and serious disease."—*The Medical Counsellor.*

"The latest and best treatise on Bright's disease."—*The Medical Era, Chicago.*

"The illustrations are numerous and good, the print large and clear, and the subject matter of the book up to date and satisfactory."—*Medical Analectic, New York.*

"Every physician should purchase and read this book. This work is written by one who has made a profound study of this departure from normal physiological action. It is simple and interesting in diction, and thorough in detail."—*The Medical World, Phila.*

"The author, besides taking advantage of the immense bibliography of this and allied subjects, is also an original investigator, and thus presents in one volume, of convenient size and brevity, the result or fruit, so to speak, of the labours of many men."—*The American Medical Visitor, Chicago.*

"The author has given the profession a most excellent work on this and allied conditions of diseased kidney. The text is clear and fully illustrated, and the treatment is well presented." — *The Richmond Southern Clinic.*

"This work is a substantial thing, and he who wants to know more of diseases of the kidneys will do well to consult it."—*American Eclectic Medical Journal.*

"We can commend this book as a well written exposition on modern ideas respecting Bright's disease. It combines the clinical and didactic aspects of the malady in well-considered relations, and will repay complete perusal."—*New York Epitome.*

Ninth Year. 8vo., Cloth Gilt, 716 pages, Illustrated. Price, 6/6, post free.

THE MEDICAL ANNUAL, 1891. Uniform in arrangement with former issues, containing ENTIRELY NEW MATTER, including important Original Contributions, designed to show the latest views and Treatment of every form of Disease, with a Synopsis of Therapeutic Agents, their Clinical Indications, and a copious Index to the entire work.

Editors and Contributors.

H. W. Allingham, F.R.C.S.; Lennox Browne, F.R.C.S.; Henry Dwight Chapin, M.A., M.D.; Samuel Craddock, M.R.C.S.; Surgeon-Major Crombie, M.D., L.R.C.S.; D. S. Davies, M.B., Lond., D.P.H., Cantab.; Prof. Dujardin-Beaumetz, M.D.; W. H. Elam, F.R.C.S.; E. Hurry Fenwick, F.R.C.S.; E. Long Fox, M.A., M.D., Oxon. F.R.C.P.; J. Dundas Grant, M.A., M.D., F.R.C.S.; F. de Havilland Hall, M.D., F.R.C.P.; G. M. Hammond, A.M., M.D.; Jonathan Hutchinson, Junr., F.R.C.S. Frank W. Jackson, M.D.; W. Allan Jamieson, M.D., F.R.C.P.; Robert Jones, F.R.C.S.E.; W. Lang, F.R.C.S.; James R. Leaming, M.D.; A. H. N. Lewers, M.D., Lond., M.R.C.P.; A. A. Liebeault, M.D.; Greville Macdonald, M.D.; R. Milne Murray, M.A., M.B., F.R.C.P.E.; W. J. Penny, F.R.C.S.; A. Mayo Robson, F.R.C.S.; A. D. Rockwell, M.D.; Robert Saundby, M D., F.R.C.P.; R. Shingleton Smith, M D., B.Sc., F.R.C.P., Lond.; J. W. Taylor, F.R.C.S.; Hugh Owen Thomas, M.R.C.S.; C. Lloyd Tuckey, M.D.; Lionel A. Weatherly, M.D., M.R.C.S; F. J. Wethered, M.D., Lond., M.R.C.P.; P. Watson Williams, M.B., Lond.; Percy Wilde, M.D.; Graham Wills, M.D.

RECENT MEDICAL WORKS

Published by John Wright & Co., Bristol.

Third Edition, now in the Press. Will be ready during the year. 10/6 post free. A New and Revised Edition brought up to date, with Numerous New Illustrations.

SURGICAL HANDICRAFT: A MANUAL OF SURGICAL MANIPULATIONS, MINOR SURGERY, Etc. For the use of General Practitioners, House Surgeons, Students and Surgical Dressers. By WALTER PYE, F.R.C.S., Surgeon to St. Mary's Hospital, and the Victoria Hospital for Sick Children; late Examiner in Surgery at the University of Glasgow. With Special Chapters on AURAL SURGERY, TEETH EXTRACTION, ANÆSTHETICS, &c., By Messrs. FIELD, HOWARD HAYWARD and MILLS.

Large 8vo, Cloth gilt, 7/6, post free. 80 Illustrations.

THE SURGICAL TREATMENT OF THE COMMON DEFORMITIES OF CHILDREN. By WALTER PYE, F.R.C.S., Surgeon and Orthopædic Surgeon to St. Mary's Hospital, and Surgeon to the Victoria Hospital for Children. Author of "*Surgical Handicraft.*" A description of the practical surgical treatment of the more common deformities of infancy and childhood.

"The Book can be safely recommended."—*British Medical Journal.*

"May be profitably studied by those interested in this branch of Surgery."—*The Practitioner.*

Fourth Edition, 8th Thousand. Revised and Enlarged. 2/-, post free. Small 8vo, Pocket Size. Cloth. Upwards of 80 Illustrations.

ELEMENTARY BANDAGING AND SURGICAL DRESSING: with Directions Concerning the IMMEDIATE TREATMENT OF CASES OF EMERGENCY. For Students, Dressers, Nurses, Ambulance Societies, &c., &c. Mostly condensed from "*Pye's Surgical Handicraft.*" By WALTER PYE, F.R.C.S., Surgeon to St. Mary's Hospital; late Examiner in Surgery at the University of Glasgow, &c. Has been adopted by the St. John Ambulance Association.

RECENT MEDICAL WORKS
Published by John Wright & Co., Bristol.

400 pages, 8vo, with over 60 Illustrations. Price 7/6, post free.

LECTURES ON MASSAGE AND ELECTRICITY IN THE CURATIVE TREATMENT OF DISEASE. By Thomas Stretch Dowse, M.D., F.R.C.P. Edin. Formerly Phys. Supt. Central London Sick Asylum, Assoc. Member Neurological Soc., New York.

"Dr. Dowse is to be congratulated on the production of a work which brings the subject of masso-therapeutics up to date."—*Glasgow Medical Journal.*

"The Drawings are well done, clear, and explanatory, and add greatly to the value of the work."—*New York Medical Journal.*

Cloth 8vo, with 39 Original Diagrams. Price 3/6, post free.

OPHTHALMOLOGICAL PRISMS: AND THE DECENTERING OF LENSES. A Practical Guide to the Uses, Numeration, and Construction of Prisms and Prismatic Combinations, and the Centering of Spectacle Lenses. By Ernest E. Maddox, M.D., Ophthalmic Surgeon, New Town Dispensary, late Syme Surgical Fellow, Edinburgh.

"We think there are few Ophthalmic Surgeons who will not learn something by its perusal."—*Ophthalmic Review.*

"From minor works we would select for special mention a little book on *The Clinical Use of Prisms*, by Mr. Ernest Maddox."—*British Medical Journal.*

8vo, Cloth, 2/6 post free. Second Edition, Enlarged.

THE WORKHOUSE AND ITS MEDICAL OFFICER. By Alfred Sheen, M.D. (St. And.), M.R.C.S.E., D.P.H. Cantab.; Senior Surgeon Glamorganshire and Monmouthshire Infirmary, Cardiff; Surgeon Cardiff Workhouse and Hospital.

"Full of useful Directions for Medical Men."—*Scotsman.*

Crown 8vo, thick Paper Covers, 1/6; or Cloth, 2/6, post free.

OUR BABY; A BOOK FOR MOTHERS AND NURSES. By Mrs. Langton Hewer, Diplomée Obstetrical Society, London; late Hospital Sister. Author of "*Antiseptic Nursing.*" The Medical Chapters have been specially written for the Book, and the whole has had the advantage of being Revised by a London Physician,

RECENT MEDICAL WORKS
Published by John Wright & Co., Bristol.

Samples Free on Application. Fiftieth Thousand.

WRIGHT'S REGISTERED POCKET CHARTS for Bedside Case Taking. Designed by Robert Simpson, L.R.C.P., L.R.C.S.   Price List:—

50 Charts, folded for Pocket Case	2/6
Ditto on Cards, Eyeleted Flat	6/-
1 Pocket Case, limp roan, to hold 58 Charts, complete with Indelible Ink Pencil	2/6
1 Guard Book, half bound, gold lettered, to hold 200 complete Charts	6/-
50 Charts, 1 Guard Book, and 1 Pocket Case, complete	10/6

Chart Holders, for hanging at Bed heads, 1/- each, or 9s. per doz.
Special Quotations to Hospitals taking not less than 250 Charts, unfolded.

REGISTERED COMBINATION TEMPERATURE AND DIET CHARTS, with Clinical Diagrams. Specially Arranged for Hospital Use. Designed by Robert Simpson, L.R.C.P., L.R.C.S. The Success of the Registered Pocket Charts has led to a desire for one upon somewhat similar lines but a larger scale for Hospital use. This is now issued on a sheet 15" × 10", and contains on the front a somewhat improved Temperature Chart for four weeks, a Diet Chart, Instructions to Nurses, etc., and at back two full length Clinical figures, including Outlines and Skeleton (front and back), with Viscera printed in red ink, and separate outlines of head. These figures will be found useful in the Treatment of Cases, and will be of special value in Hospitals where Clinical Teaching is carried on, as by their means the teacher is enabled more clearly to demonstrate any diseased condition of the organs, also the position of fractures or areas of local paralysis, etc., etc.   Price List:—

	per 1000	500	250	100	50	less.
Charts, complete, with figures on back in two colours	36/-	20/-	11/-	5/-	3/-	1½d. each.
Ditto without figures, one side only	22/-	12/6	7/-	3/-	2/-	1d. ,,
Clinical Figures only, two colours	26/-	15/-	8/-	3/6	2/6	1d. ,,

www.ingramcontent.com/pod-product-compliance
Lightning Source LLC
Chambersburg PA
CBHW032223230426
43666CB00033B/834